Macmillan

ENCYCLOPEDIA OF SCIENCE

1

Matter and Energy

Peter Lafferty

Macmillan Publishing Company
New York

Maxwell Macmillan International Publishing Group
New York Oxford Singapore Sydney

Board of Advisors
Linda K. Baloun, B.S., Science Teacher, Highmore, SD

Merriley Borell, Ph.D., Consultant, Contemporary Issues in Science and Medicine, Piedmont, CA

Maura C. Flannery, Ph.D., Associate Professor of Biology, St. John's University, Jamaica, NY

Owen Gingerich, Ph.D., Professor of Astronomy and History of Science, Harvard-Smithsonian Center for Astrophysics, Cambridge, MA

MaryLouise Kleman, M.A., Science Teacher, Baltimore County, MD

Irwin Tobias, Ph.D., Professor of Chemistry, Rutgers University, New Brunswick, NJ

Bibliographer
Sondra F. Nachbar, M.L.S., Librarian-in-Charge, The Bronx High School of Science, Bronx, NY

Advisory Series Editor
Robin Kerrod

Consultant
Dr. B. R. Orton

U.S. Editorial and Design Staff
Jean Paradise, Editorial Director
Richard Hantula, Senior Editor
Mary Albanese, Design Director
Zelda Haber, Design Consultant
Norman Dane, Cover Design

Andromeda Staff
Caroline Sheldrick, Editor
David West/Children's Book Design, Designer
Alison Renney, Picture Research
Steve Elliott, Production
John Ridgeway, Art Director
Lawrence Clarke, Project Director

Front Cover: Laser welding *(Meta Machines Ltd.)*

Published by:
Macmillan Publishing Company
A Division of Macmillan, Inc.
866 Third Avenue, New York, NY 10022

Collier Macmillan Canada, Inc.
1200 Eglinton Avenue East, Suite 200
Don Mills, Ontario M3C 3N1

Planned and produced by Andromeda Oxford Ltd.

Library of Congress Cataloging-in-Publication Data

Macmillan encyclopedia of science.
p. cm.
Includes bibliographical references and index.
Summary: An encyclopedia of science and technology, covering such areas as the Earth, the ocean, plants and animals, medicine, agriculture, manufacturing, and transportation.
ISBN 0-02-941346-X (set)
1. Science–Encyclopedias, Juvenile. 2. Engineering–Encyclopedias, Juvenile. 3. Technology–Encyclopedias, Juvenile. [1. Science–Encyclopedias. 2. Technology–Encyclopedias.]
I. Macmillan Publishing Company 90-19940
Q121.M27 1991 CIP
503 – dc20 AC

Volumes of the *Macmillan Encyclopedia of Science*
1 *Matter and Energy* ISBN 0-02-941141-6
2 *The Heavens* ISBN 0-02-941142-4
3 *The Earth* ISBN 0-02-941143-2
4 *Life on Earth* ISBN 0-02-941144-0
5 *Plants and Animals* ISBN 0-02-941145-9
6 *Body and Health* ISBN 0-02-941146-7
7 *The Environment* ISBN 0-02-941147-5
8 *Industry* ISBN 0-02-941341-9
9 *Fuel and Power* ISBN 0-02-941342-7
10 *Transportation* ISBN 0-02-941343-5
11 *Communication* ISBN 0-02-941344-3
12 *Tools and Tomorrow* ISBN 0-02-941345-1

Printed in the United States of America

Introduction

This volume tells about basic aspects of physics and chemistry. Much of the information is presented in photographs, drawings, and diagrams. They are well worth your attention.

To learn about a specific topic, start by consulting the Index at the end of the book. You can find all the references in the encyclopedia to the topic by turning to the final Index, covering all 12 volumes, located in Volume 12.

If you come across an unfamiliar word while using this book, the Glossary may be of help. A list of key abbreviations can be found on page 87, and a handy table of the chemical elements and their symbols on page 31. If you want to learn more about the subjects covered in the book, the Further Reading section is a good place to begin.

Scientists tend to express measurements in units belonging to the "International System," which incorporates metric units. This encyclopedia accordingly uses metric units (with American equivalents also given in the main text). In illustrations, to save space, numbers are sometimes presented in "exponential" form, such as 10^9. More information on numbers is provided on page 87.

Contents

Part One

The nature of matter

Everything in the world about us and in the Universe as a whole is made up of stuff we call matter. Matter appears in an infinite variety of different guises: pebbles on the beach, dew on the grass, clouds in the sky, wildlife on the prairies, stars in the heavens.

However, the many millions of different substances that exist are in fact made up of only about 90 simple "building blocks." We call them the chemical elements. These elements exist as tiny particles called atoms, usually linked together in groups in the form of molecules. Atoms and molecules are always on the move.

In their turn atoms contain even tinier particles, such as protons and electrons. It is the number and arrangement of these particles inside atoms and molecules that determine the chemical and physical properties of matter.

◀ Menthol crystals used in cough medicine. The structure of matter is revealed using many different techniques. This photograph was taken in polarized light. This effect is used in polarizing sunglasses.

Solids, liquids, and gases

The matter of the everyday world comes in three familiar states called solids, liquids, and gases. Solids, such as a piece of steel, have a fixed shape which stays the same at ordinary temperatures. Liquids, like water and milk, have no fixed shape. They take on the shape of the container that holds them. Gases, such as the air in a toy balloon, are also shapeless and they fill the whole of their container. These properties of solids, liquids, and gases come about because their atoms or molecules – the tiny particles that make them up – are held together with differing strengths. Changes from one state to another occur when the atomic or molecular arrangement changes, usually because of a change in temperature.

Spot facts

- *The densest material in the Universe occurs inside the smallest stars, called neutron stars. A pinhead made of this material would weigh more than two supertankers.*
- *Solids are not always denser than liquids. Ice is less dense than water. This is why icebergs float in the sea. Pumice, a kind of lava, can float on water too – the only rock to do so.*
- *Smoke and fog are mixtures of small particles and gases. These mixtures are called aerosols.*
- *The molecules in air at ordinary temperatures are moving at about 10 times the speed of the winds in a hurricane.*

► The water in the lake is a liquid. Ice and snow – water in the solid state – can be seen on the mountain slopes. The gaseous form of water – water vapor – is invisible but it is present in the air. The clouds on the mountaintop are formed when water vapor condenses into water droplets.

Moving molecules

A solid is dense and rigid because its molecules are bound tightly in place. The attractive force between the molecules is strong. Because of this, the molecules are held in a more or less fixed position. However, they can vibrate, or move back and forth slightly, in all three directions. Their vibrations increase in speed as the solid is heated. If a solid such as a metal is heated sufficiently, the molecules vibrate so much that the attractive forces cannot hold the metal atoms in their rigid structure.

In a liquid, the attraction between molecules is weaker. They are able to move about the way a person can move about in a crowd. This leaves empty spaces into which other molecules can move. This movement of molecules enables the liquid to flow easily.

In gases, the molecules move about completely freely. Their speed increases as the gas is heated. This way of looking at solids, liquids, and gases is called the kinetic theory.

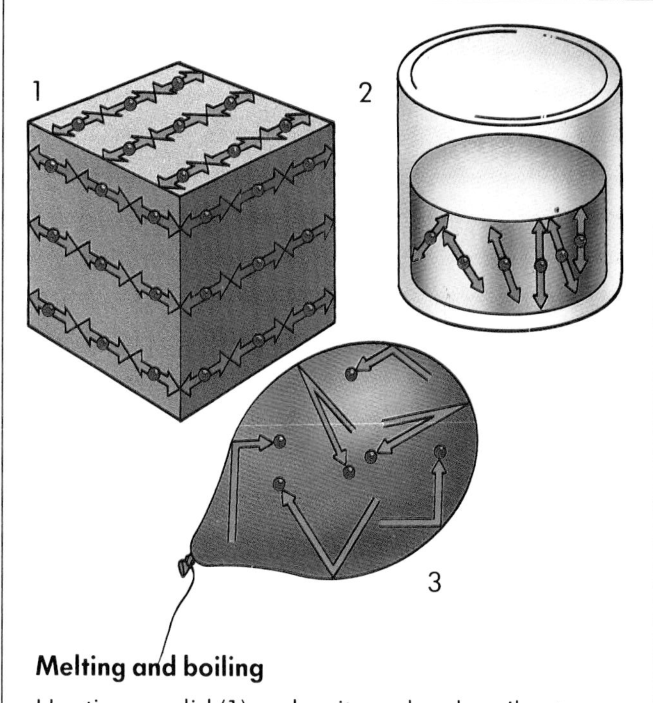

Melting and boiling

Heating a solid (1) makes its molecules vibrate more vigorously. When the temperature is high enough, the molecules give up their fixed positions. The solid melts and turns into a liquid (2). Further heating makes the molecules vibrate even more. Eventually the liquid boils and turns into a gas (3).

▼ A volcanic eruption involves matter in all three states. The volcano and the surrounding land are made of solid rock. Molten rock bubbling out is a liquid. And the fumes blown high into the air are gases.

Solids

Most solids are made up of crystals, pieces of material that have flat surfaces with straight edges. Salt and sugar are familiar examples of crystals. However, a microscope reveals that other solids, such as steel and copper, are also made up of small crystals. X-rays show that crystals are composed of a regular arrangement of atoms. They are spaced very close together – only a few tenths of a nanometer apart. (A nanometer is about one billionth of a yard.) The geometric shape of a crystal reflects the regular arrangement of its atoms.

Different arrangements

Many properties of a solid depend upon its atomic or molecular arrangement. An interesting example is the element carbon, which occurs in two very different forms. One form, diamond, is a very hard material. Its hardness results from the very strong chemical bonds that form between the atoms in a diamond crystal. The other form of carbon, graphite, is one of the softest substances. It has a structure in which the carbon atoms lie in layers. There are only weak links, or bonds, between the

Crystal shapes

A crystal has flat faces at angles to each other. The shape of the faces and the angles between them depend on the arrangement of the atoms in its molecules. In different crystals of the same mineral, the angles between faces are always the same (right). The color of a crystal also depends upon its molecular makeup. Quartz (below) is clear, whereas feldspar (below right) is pearly.

layers. As a result, graphite is a relatively weak material, used as a lubricant.

Many other qualities of a solid also depend upon the strength of the forces between its atoms or molecules. The melting temperature is an example. A solid with strong bonds between its atoms needs lots of heat energy to melt it. The amount a solid expands when it is heated also depends upon the strength of the forces between its atoms or molecules. As a solid is heated, the atoms or molecules vibrate more rapidly. They are able to move apart slightly if the forces holding them together are not too strong. So the average distance between the molecules increases, and the material expands.

Elasticity also depends upon the forces between atoms or molecules. If a weight is hung from a wire, the wire stretches slightly. As long as the weight is not too heavy, the wire returns to its original length when the weight is removed. This is called elastic stretching. Materials that stretch least are those with strong forces between their atoms. As long as the weight is not too heavy, the force between the atoms pulls them together again when the weight is removed. But if too great a weight is used, the wire does not return to its original length when the weight is removed. It remains stretched. In this case, the bonds between the atoms in the wire are permanently lengthened.

▼ Large structures such as bridges have to built of strong materials. They have to be strong enough to support the structure itself as well as any extra weight it has to carry. They also have to be designed to withstand the effects of expansion in hot weather.

An expansion joint in a railroad line.

When railroad tracks are laid, small gaps are left between the ends of the rails. When the tracks become hot in summer, the metal rails expand and fill the gaps. Without expansion joints, the expanding track would buckle.

Liquids

The forces between molecules in a liquid are weaker than those in a solid. Nevertheless, these forces produce effects that can easily be seen in everyday life. One effect is called surface tension. It causes the surface of a liquid to behave as if it is covered with a thin rubber skin. This tension causes the roundness of small drops of liquid, and lets small insects walk on water. The tension arises because the molecules at the liquid surface are pulled toward the center of the liquid by the attracting force of the other molecules.

Surface tension makes water climb up a fine glass tube that is dipped into water. This effect is called capillarity. It happens because the attractive forces between the glass molecules and the water molecules are greater than those between the water molecules themselves. The surface of the water is pulled upward at the edges of the tube, creating a curved surface where the water wets the glass. Some liquids, such as mercury, drop down small tubes. This happens because the mercury does not "wet" the glass. Its molecules attract one another more than they attract the glass molecules.

▼ Sugary syrup running off a spoon shows all the features of a viscous liquid. It is thick and sticky, and it pours only with difficulty.

▼ A pond skater walks on the surface of a pond. At the surface of a liquid, forces pull the molecules inward. This surface tension is enough to support the insect.

Liquids flow easily because there are only weak forces between the molecules of a liquid. However, there is some resistance to flow caused by these forces. This resistance is called viscosity. A liquid with a high viscosity – a viscous liquid – flows only slowly, like syrup.

Some liquids are good solvents – that is, other substances can dissolve in them to form a solution. Water is a good solvent for many substances, such as salt and sugar, which are said to be soluble in water. Gases like oxygen and carbon dioxide can also dissolve in water. A third type of solution is formed when two liquids mix together. For example, water and alcohol mix to form such a liquid solution.

The amount of a substance that dissolves in a given amount of a solution is called its solubility. For a solid, solubility depends on the temperature. Sugar is much more soluble in hot water than in cold. For a gas, solubility also depends on pressure. The higher the pressure, the greater the solubility of the gas.

▼ A dam stores energy in a vast lake of water. The pressure of the water increases with depth, and is greatest near the base of the dam. For this reason, the dam is much thicker at the base than at the top. The energy is carried in flowing water, which spins turbines to generate electricity.

▼ The weight of the ice in a glacier causes it to flow downhill, like a very viscous liquid. A thin film of water helps the glacier to slip over the rocks.

Gases

A gas consists of molecules traveling at high speed. Each cubic centimeter of air we breathe (about 1/16 cu. in.) contains about 20 quintillion (20 followed by 18 zeros) molecules at ordinary temperature and pressure. The molecules dart about at 450 meters (1,500 ft.) per second. Each molecule travels only a short distance before it collides with another – about one ten-millionth of a meter (about a quarter of a millionth of an inch).

Minute molecules

Molecules of gas are too small to be seen directly. However, when smoke particles suspended in air are examined with a microscope, they are seen to make small irregular movements. This motion is called Brownian motion after the British scientist Robert Brown, who first saw it in 1827. The movements of the smoke particles are caused by air molecules constantly hitting them.

▼ A hot-air balloon rises because the heated air inside it is lighter than the unheated and denser gas surrounding it. Heated gas becomes less dense because its molecules move more rapidly, so that the average distance between them increases.

Gas can be compressed very easily. This is because there is a relatively large amount of space between gas molecules, so it is easy to squeeze them closer together. One of the first scientists to study the connection between the volume of a gas and the pressure on it was the Irish scientist Robert Boyle. He showed in 1662 that if the pressure on a gas is doubled, the volume is halved, as long as the temperature does not change. In general, the product of the pressure and the volume is constant for a given mass of gas. This is called Boyle's law.

Another important law describes the way the volume of a gas changes when it is heated or cooled. This is called Charles's law. It states that if the volume of a gas is known at 0°C (32°F), the volume increases or decreases by 1/273 of this value for every degree (Celsius) rise or fall in temperature, as long as the pressure does not change. Thus at a temperature of −273°C (about −460°F), a gas will have zero volume. This temperature, called absolute zero, must therefore be the lowest possible temperature.

Inside the Sun

The matter at the center of the Sun can reach temperatures of about 14 million degrees Celsius (25 million °F). At these temperatures, the atoms of the material inside a star such as the Sun are ripped apart. An extremely hot gas made up of subatomic particles called protons and electrons is formed. The gas is called a plasma.

▼ A car being tested in a wind tunnel to see how well it slips through the air. Streamlining the car helps to reduce air resistance.

▼ The winds that push a sailing ship along are caused by pressure differences in the atmosphere. Over hot regions the air rises, creating an area of low pressure below it. Air from cooler regions flows into this low-pressure zone as wind.

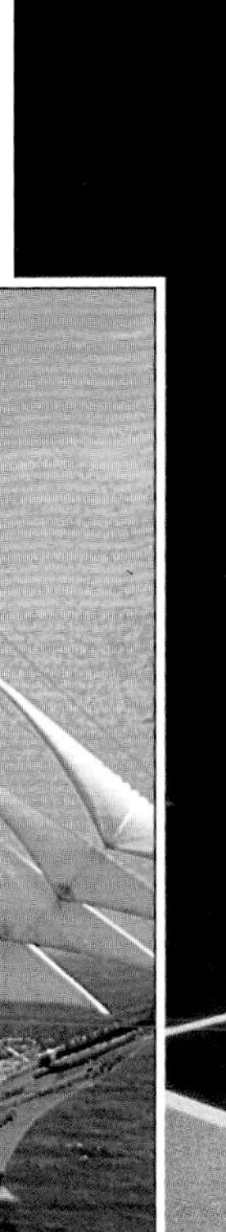

Changes of state

The temperature of a substance is a measure of the average energy of its molecules. The higher the temperature, the greater the average energy. Not every molecule has the same energy, however. As the temperature of a crystalline solid rises, the number of molecules with enough energy to move freely increases and the crystal starts to melt. As more heat is added, even more molecules move away from their fixed positions. While this takes place the temperature of the substance remains constant until it is completely melted. The solid has changed state into a liquid.

The heat energy required to melt a solid completely is called the latent heat of fusion. This heat may be supplied by a candle flame or an electric cooker, for example, or it may come from the solid's surroundings. This explains why a block of ice cools its surroundings. Heat is extracted from the surroundings in order to melt the ice.

A similar effect takes place when a liquid evaporates, or turns into a gas. Energy is needed for the molecules of the liquid to escape from their neighbors and form a gas. When water evaporates from your skin, for example, the skin feels cooler because it has lost heat to the evaporating water.

◀ The energy needed to melt steel during laser welding is supplied by a powerful beam of light. The laser beam is concentrated on a very small area at a time, allowing precise cutting and welding.

Much more energy is needed for molecules in a liquid to form a gas than for molecules of a solid to form a liquid. For example, at room temperature, only a tiny fraction of the water molecules in a bowl have enough energy to evaporate. However, within a few days all the water evaporates away because the water molecules diffuse away from the bowl. Eventually all the molecules acquire sufficient energy to escape from the liquid.

The process of evaporation can be speeded up by heating a liquid. When the liquid is hot enough, it boils. But even at the boiling point, more heat energy is needed to convert all of the liquid into a gas, or vapor. This energy is called the latent heat of vaporization. The bubbles in a pan of boiling water, for example, are bubbles of water vapor that form at the bottom of the pan and then rise to the surface.

▲ A skater moves on a thin film of water formed when pressure melts the ice under the skates.

▼ Mist is caused when a rise in temperature evaporates water to form a gas, or vapor. This is evaporation. When the vapor meets a layer of colder air, it forms tiny droplets of mist. This is condensation.

▼ Clouds of mist form when dry ice – solid carbon dioxide – is thrown onto a stage. The change of a solid into a gas is called sublimation.

Atoms and molecules

Spot facts

- *A gold wedding ring contains about 10 sextillion atoms. About 100 billion atoms would fit on the period at the end of this sentence.*
- *Atoms are mostly empty space. If the nucleus were the size of a tennis ball, the nearest electron would be a kilometer (over half a mile) away.*
- *A proton in an atom is about 2,000 times heavier than an electron.*
- *Most molecules are made up of small numbers of atoms, but many contain more. Aspirin molecules contain 21 atoms. Rubber molecules may have up to 65,000 atoms. Sugar may have up to 150,000 atoms in its molecules.*

Atoms and molecules are very small, yet scientists have discovered many things about them. They have found that there are even smaller particles inside atoms. Most of the mass of an atom is concentrated at its center, in the nucleus. The nucleus is made up of particles called protons and neutrons. Even smaller particles called electrons circle around the nucleus. The number of electrons and the way they are arranged around the nucleus are among the most important features of an atom. They determine how atoms join together to form different chemical substances. Molecules are formed when atoms join together. Simple molecules may have only two atoms. Complex molecules, such as those found in living things, may have many thousands of atoms.

▶ Particles much smaller than an atom can be detected in a bubble chamber. As they travel through the chamber, they leave behind trails of bubbles which are then photographed. By studying these tracks, scientists are able to identify the particles.

Inside the atom

The idea that matter is made up of atoms is a very old one. The ancient Greek thinker Democritus, who lived about 400 BC, argued that all matter was made up of very small particles. These particles were, he thought, indestructible. In 1802 an English scientist, John Dalton, provided the first evidence that atoms were real. He agreed with Democritus that atoms were indestructible and indivisible. They were like small billiard balls, Dalton thought. He was able to measure the weights of some atoms and explain how atoms joined together to form molecules. His atomic theory was accepted for many years.

Then in 1897 another English scientist, J. J. (Joseph John) Thomson, proposed that atoms were not indivisible. They contained even smaller particles, called electrons. Thomson thought electrons were scattered throughout the atom, like raisins in a fruitcake.

The center of an atom

Additional insight into the atom was provided in 1911 by Ernest Rutherford, a New Zealander who did most of his scientific work in England. Rutherford bombarded a thin piece of gold with alpha particles. Alpha particles are rays given off by certain radioactive materials, such as radium. He was amazed to find that a few alpha particles bounced off the gold rather than going straight through.

This could happen only if most of the mass in the atom were concentrated into a very small region at its center. We now call this very dense region the nucleus. The electrons circled around the nucleus in a sort of "electron cloud," Rutherford said. In 1913 the Danish scientist Niels Bohr suggested that the electrons circled around the nucleus in a series of orbits at different distances from the nucleus, in much the same way as the planets orbit the Sun.

◀ The English scientist J. J. Thomson sitting in front of a cathode-ray tube. He used the cathode-ray tube to study electrons. From his experiments he was able to demonstrate that electrons are present in all atoms.

▼ In 1802 John Dalton pictured atoms as solid objects like billiard balls. When J. J. Thomson discovered the electron, he imagined electrons scattered through the atoms. In 1911 Ernest Rutherford discovered the nucleus and pictured electrons in orbit around it. In 1913 Niels Bohr showed these orbits are arranged in layers, or "shells."

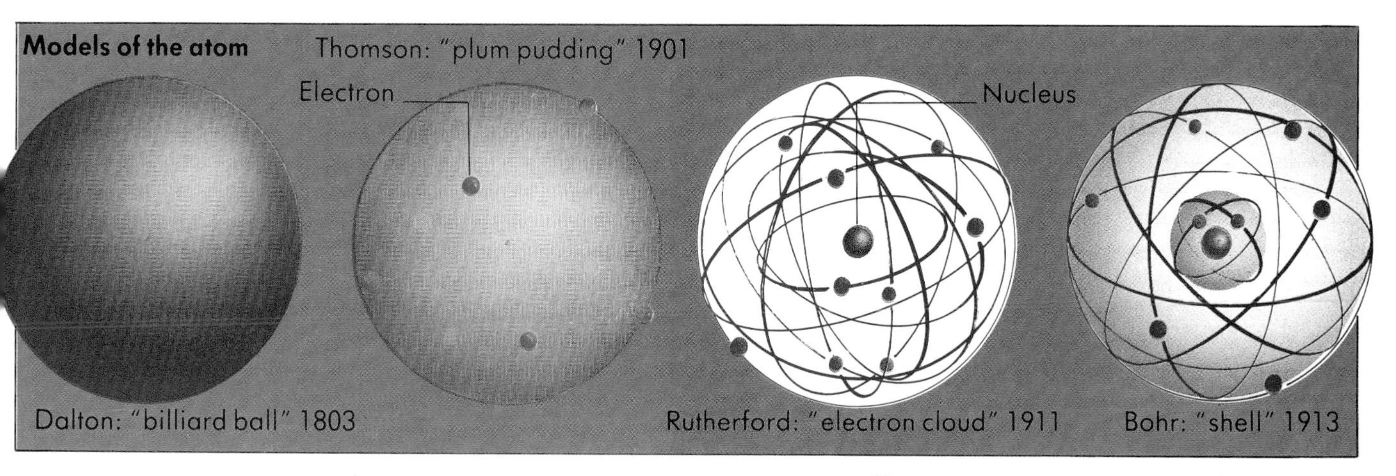

Atomic structure

We now know that the atomic nucleus contains particles called protons and neutrons. Like electrons, protons carry a tiny amount of electricity – they are said to be charged. Protons, however, carry a different kind of electricity from electrons. Scientists say that the proton has a positive charge, and the electron has a negative charge. The neutron, the other particle found in the nucleus, has no charge. It is electrically neutral.

Niels Bohr thought that electrons move in clearly defined paths, or orbits, around the nucleus. In each of these orbits, they require a certain amount of energy to keep them from the nucleus. The electrons in orbits close to the nucleus require more energy than those further out. When an electron moves from one orbit to another orbit closer to the nucleus, the electron gives off energy, which may be in the form of a small flash of light.

Wave patterns

Later scientists took Bohr's ideas into account but added the idea that electrons, and other small particles, can sometimes behave like waves. When electrons pass through a very narrow slit, for example, they appear to spread out like a wave does when going through a slit. Inside the atom, the wavelike behavior of electrons means that it is impossible to say exactly where an electron is at any time. So, instead of saying that the electrons are in precise orbits, scientists draw maps of where electrons are likely to be.

The region of space occupied by an electron is called an orbital. The simplest orbital is like a ball. It can hold two electrons. Other orbitals have more complicated shapes, such as a dumbbell and an hourglass. By discovering how electrons fit into an atom's orbitals, scientists can explain its chemical properties.

Inside the nucleus

Within the nucleus of an atom there are positively charged protons and neutrons, which have no electrical charge. The number of protons equals the number of negatively charged electrons circling the nucleus, so that the atom has no overall electric charge. Modern experiments use beams of high-energy electrons to probe inside the protons and neutrons themselves. These experiments show that they contain dotlike particles called quarks.

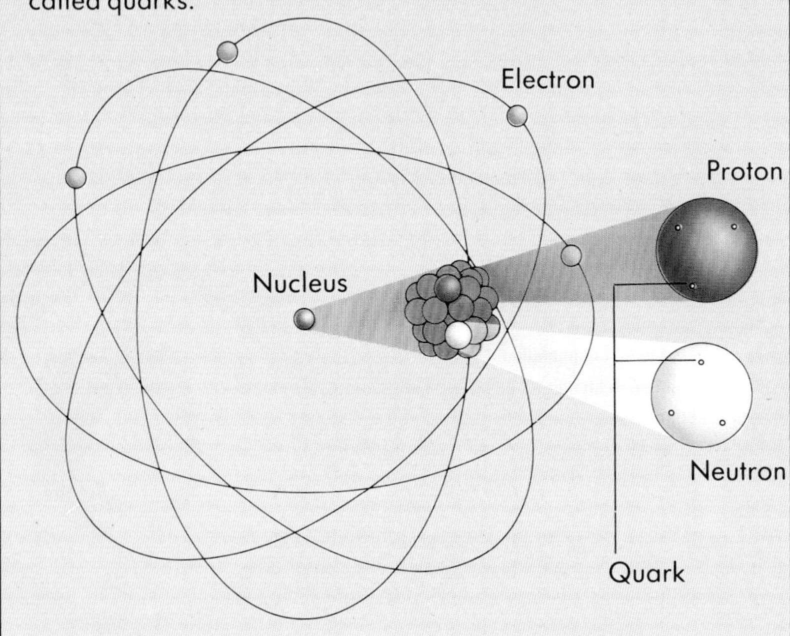

▼ A group of uranium atoms. The picture was taken using a scanning electron microscope. In this instrument a fine beam of electrons is swept, or scanned, across the object being examined. Some of the electrons are reflected off the object and used to form the picture. The latest electronic imaging techniques can be used to photograph individual atoms which are only 30 billionths of a centimeter (75 billionths of an inch) across.

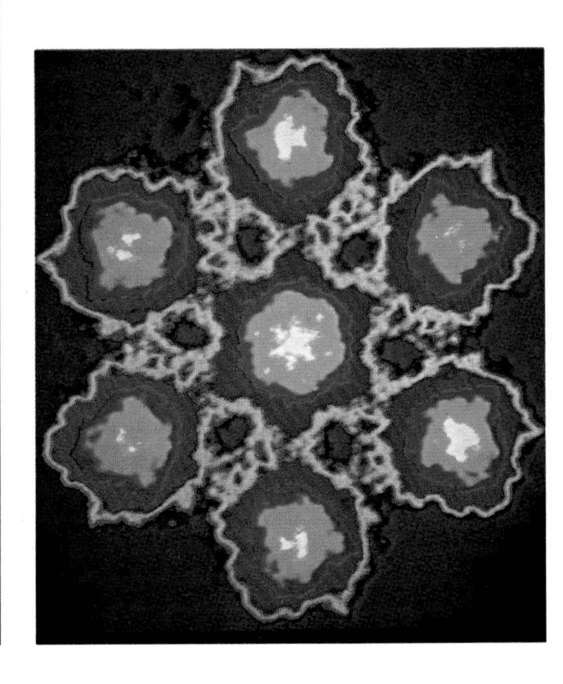

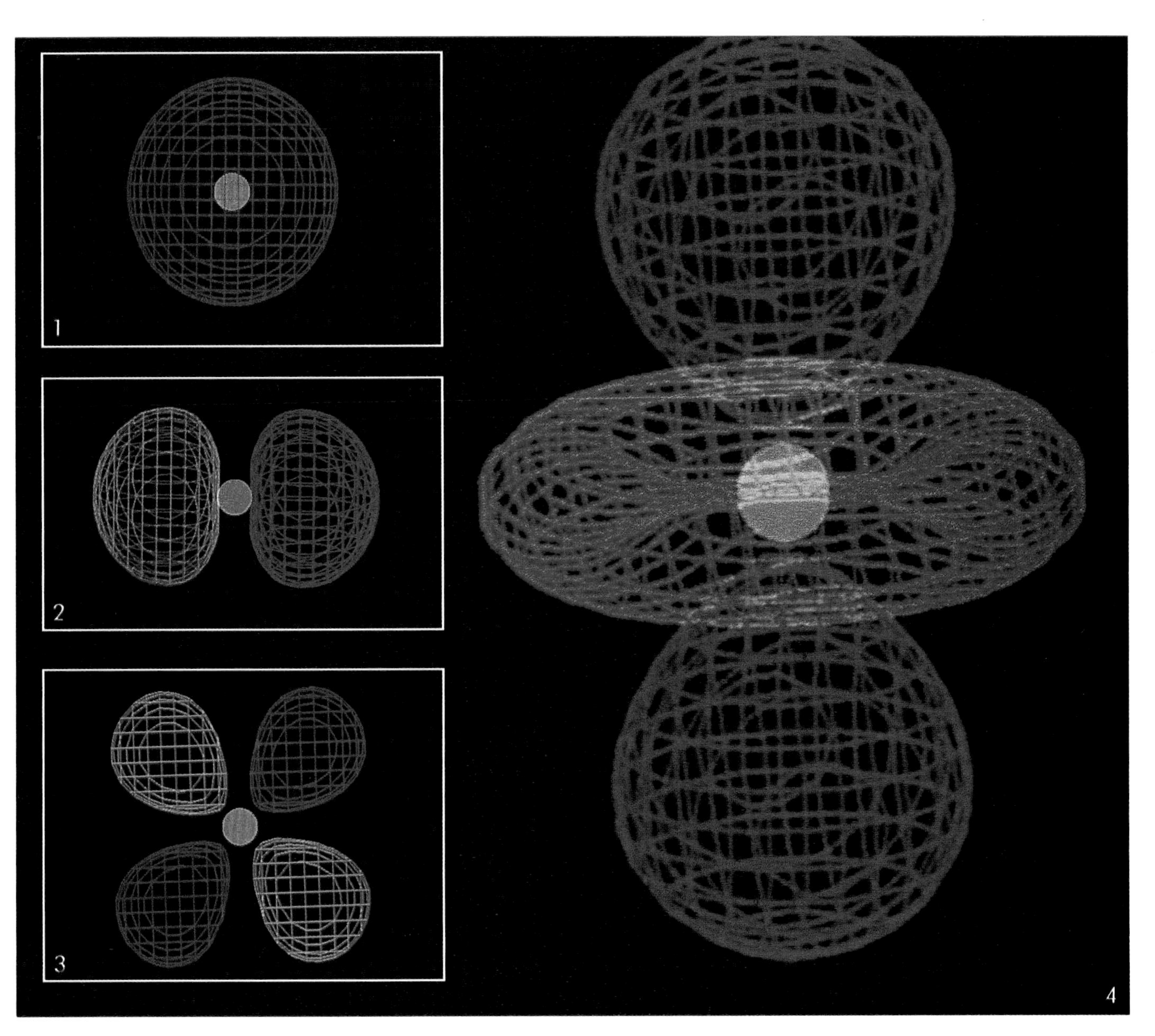

▲ In the wave picture of the atom, scientists draw "orbitals." Orbitals are regions of space where the electron is likely to be, although its exact location cannot be stated. There are four simple shapes of orbital: (1) spherical, or ball-like, (2) dumbbell, (3) four-leaf clover, (4) hourglass and ring. Other orbitals are much too complicated to draw.

▶ The particle picture of the atom has electrons circling the nucleus, like planets around the Sun, as in the atoms shown here. The first orbital out from the nucleus can hold two electrons, the second orbital can hold eight electrons, and so on. The simplest atom, hydrogen, has one electron in the first orbital. Oxygen, with a total of eight electrons, has two in the first orbital and six in the second. Magnesium has two electrons in its third orbital.

The buildup of electrons

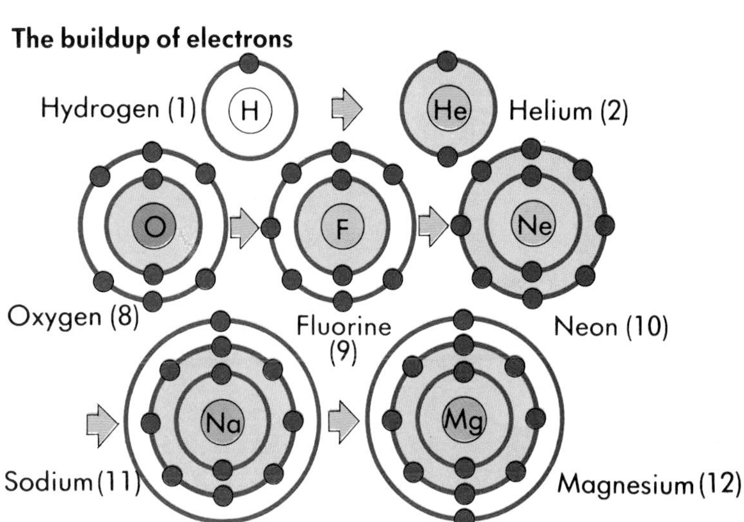

Particles galore

To study the particles that make up the nucleus of an atom, scientists use particle accelerators, popularly known as atom smashers. These are large machines in which particles are boosted, or accelerated, to high speeds. The accelerated particles collide with the nuclei in a target, producing a shower of other particles, which have very short lifetimes. The tracks of these particles are recorded using many types of detectors, such as a bubble chamber. As the particles travel through the liquid in the chamber, they leave tracks of bubbles behind. Magnets around the chamber are used to bend the tracks of electrically charged particles. By seeing how the particles are affected by the magnets, scientists are able to identify them.

These experiments reveal that there are many different subatomic particles. There are heavy particles, such as the pion and the kaon. Like the neutron and the proton, these particles are made up of quarks. The proton and neutron each have three quarks. The pion and kaon are made up of two quarks. There are also light particles, such as the neutrino. This particle appears to have no mass at all. Neutrinos are not made up of quarks. Neither are electrons and other slightly heavier particles called the tau and the muon. The story is not yet over, because scientists still have more things to discover about particles.

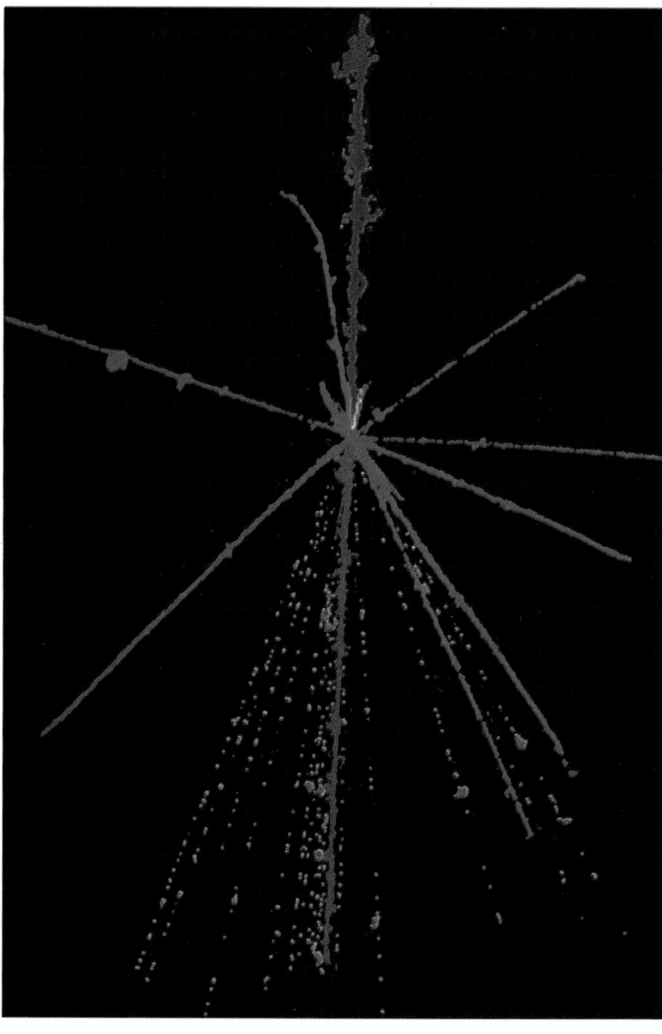

▲ Tracks of cosmic rays. They are high-energy particles that reach Earth from outer space. They have been captured in a photographic emulsion. The resulting picture was later colored to identify the tracks. A sulfur nucleus (red) has collided with a nucleus in the film to produce a spray of particles including a fluorine nucleus (green), 16 pions (yellow), and several other nuclear fragments (blue).

◀ An early type of particle detector was the cloud chamber, invented by Scottish physicist Charles Wilson in 1895. If the globe is filled with moist air and the pressure reduced suddenly, water droplets form and make particle tracks visible.

Tracking particles

In this photograph of particle tracks in a bubble chamber, colors have been added. Negative particles, called kaons, enter the picture from below. Their tracks curve to the right slightly, showing that the chamber's magnetic field deflects negative particles (shown in purple, pale blue, and green) to the right. Positive particles (orange and red) are deflected to the left.

The two spirals are caused by electrons, the only particles light enough to curl so tightly in the magnetic field. The tiny spiral is from an electron knocked from an atom in the bubble-chamber liquid. The large spiral comes from the breakup of the particle that made the pale blue track. This must have been a muon.

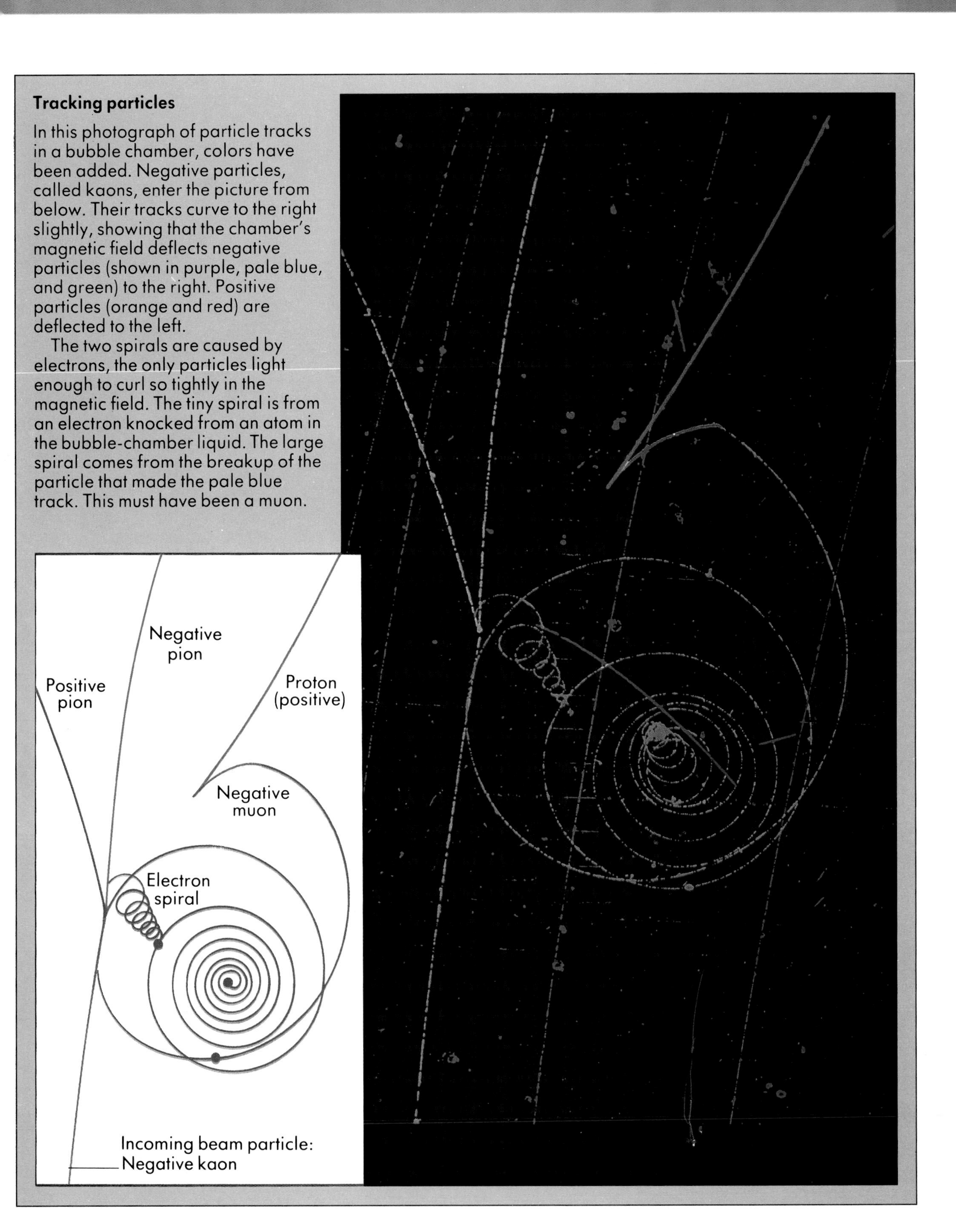

Radioactivity

All the atoms of a particular element have the same number of protons in their nuclei. But the number of neutrons may vary. Atoms that have the same number of protons, but a different number of neutrons, are called isotopes.

Nearly all chemical elements have several isotopes. There are three isotopes of hydrogen, for example. The most common isotope of hydrogen has a single proton in its nucleus. The other isotopes, called deuterium and tritium, have two and three particles in their nuclei, respectively.

Unstable isotopes

The nuclei of many isotopes are stable and unchanging. However, some isotopes are unstable. The nuclei of these isotopes break up, or decay. These isotopes are said to be radioactive. They give out energy in the form of radiation in order to become more stable. Three different types of radiation are given out by radioactive isotopes. One type is called alpha radiation. This consists of a stream of tiny particles, called alpha particles, each made up of two protons and two neutrons. The second type is called beta radiation. It consists of high-energy electrons, or beta particles. The third type of radiation is gamma radiation, which resembles high-energy X-rays.

Half-life

Not all the nuclei of a radioactive isotope decay at the same time. One nucleus may decay quickly, another may take a long time. But the decay process always proceeds at the same rate, on average. The time taken for half the nuclei in a sample to decay is always the same. This time is called the half-life of the isotope.

The half-lives of some isotopes are as short as a fraction of a second, whereas other isotopes decay much more slowly, with half-lives of millions of years. This provides a clue to which isotopes occur naturally. Those with short half-lives, compared to the age of the Earth, have long since decayed and disappeared. Some, such as carbon, take longer to decay. By measuring the amount of radioactive isotopes remaining in samples of old materials, it is possible to estimate how old they are. This process is called radiocarbon dating.

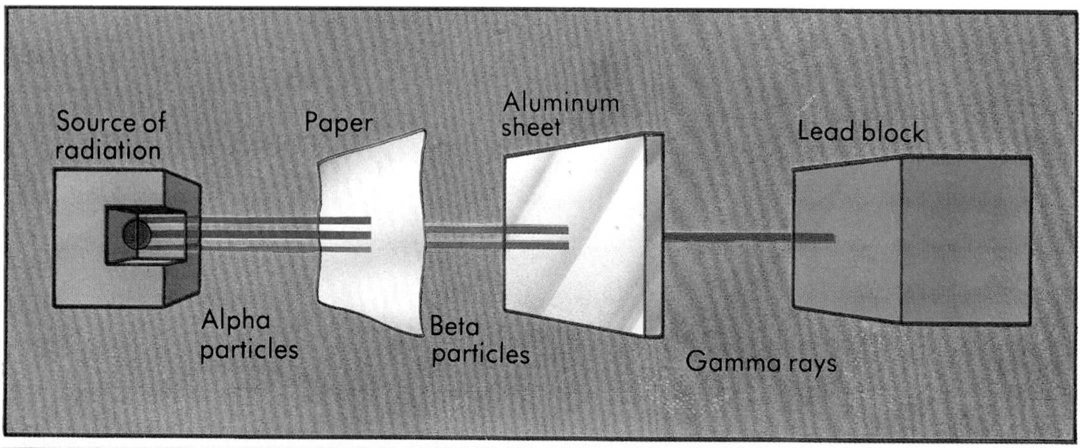

◀ The three kinds of radiation given out by a radioactive substance have different penetrating power. Alpha particles are stopped by a sheet of paper; beta particles by a thin sheet of metal; gamma rays by a thick lead block.

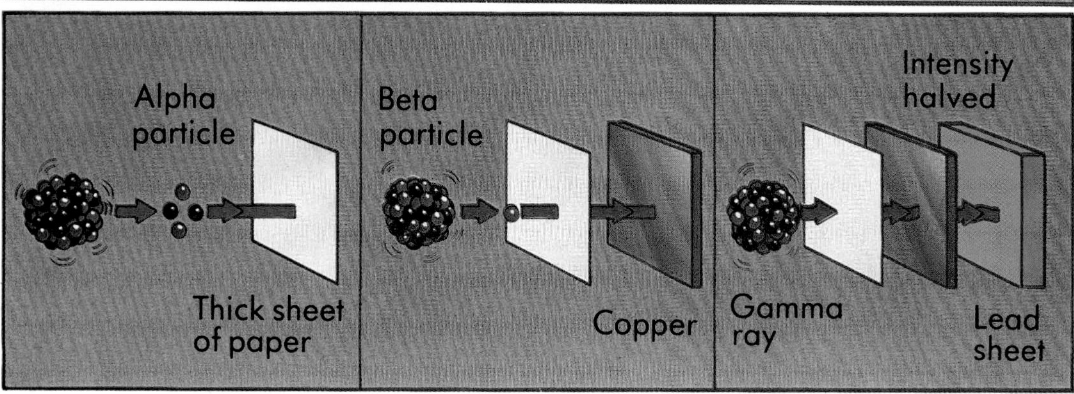

◀ Alpha particles have the lowest energy. Beta particles are more penetrating than alpha particles because they are smaller and move at higher speeds. More penetrating still are gamma rays. They are similar to X-rays but have a shorter wavelength.

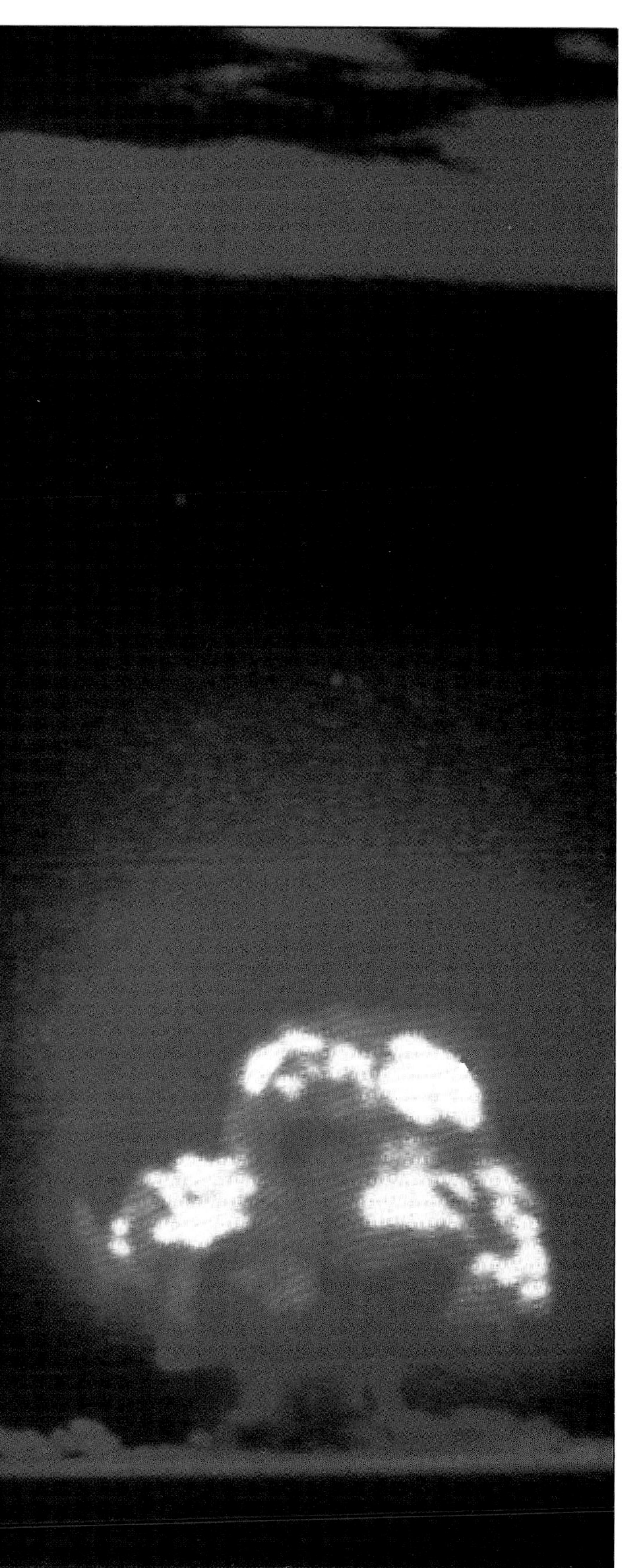

The second way to release nuclear energy is fusion. This is the joining of light nuclei, such as an isotope of hydrogen, to make heavier nuclei. Fusion is the process that produces energy inside the Sun and other stars. For many years, scientists have been trying to produce energy from hydrogen fusion here on Earth. They have been successful in producing hydrogen bombs, which release fusion energy in a sudden, violent manner. But they still have not produced fusion energy in a controlled manner.

The main problem is that a very high temperature, about 100 million degrees Celsius, is needed before hydrogen nuclei can be made to combine. At lower temperatures, their electric charges cause them to repel each other. Fusion would be a cheap source of energy because the hydrogen fuel needed could be extracted from water. However, it will be many years before fusion power is a reality.

◀ The first atomic bomb explosion, in New Mexico on July 16, 1945, worked on the principle of fission. In the more powerful hydrogen bomb, an atomic bomb produces the high temperature needed for fusion.

▼ The aim in nuclear fusion is to bring together two heavy isotopes of hydrogen, tritium and deuterium, so that they fuse, or join. The result is a helium nucleus, a neutron, and the release of energy.

Hydrogen fusion

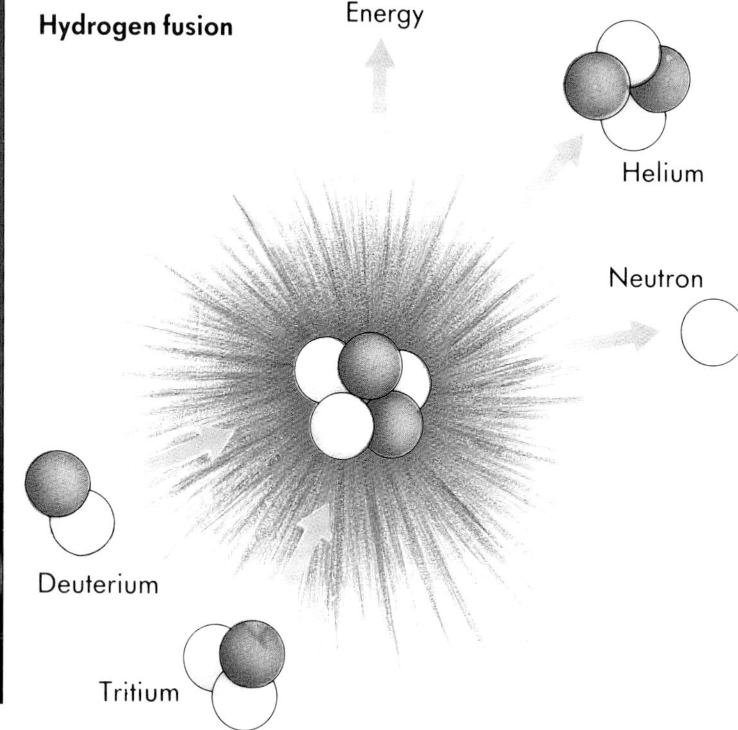

Chemical elements

Spot facts

- *Of the 92 naturally occurring chemical elements, only two are liquids at ordinary temperatures: bromine and mercury. Only 11 elements are gases. All the other elements are solids, mostly metals.*
- *The most common element in the Universe is hydrogen. Most stars are made up of nine-tenths hydrogen and one-tenth helium.*
- *The rarest naturally occurring element is astatine. It is believed that there is only a third of a gram of astatine in the whole of the Earth's crust.*

All the countless different materials that exist in the Universe consist of combinations of simple substances, called elements. The elements are the basic building blocks from which all other substances are made up. When elements combine chemically, they form compounds. Chemists explain the properties of the elements, how they combine to form compounds, and how chemical reactions occur, by looking at the way the electrons are arranged in the atoms.

▶ With their primitive apparatus, seen in this 18th-century painting, early chemists, or alchemists, could perform only simple experiments. They searched for a "philosopher's stone," which could turn common metals into gold. They also sought medicines that could prolong life forever. Despite failing in these aims, the alchemists made many chemical discoveries and laid the foundations of the science of chemistry.

Elements and compounds

Substances can be elements, compounds, or mixtures. A mixture is a substance that can be separated into different materials by means, such as filtering, that can be easily reversed. These changes are called physical changes. Seawater is a mixture. The salt and water can be separated by heating the seawater until the water has evaporated, leaving the salt behind. The change can be reversed easily by pouring the salt back into water and stirring. A mixture consists of atoms or molecules that are not connected together and so are easy to separate.

Compounds can be separated into different substances too. But the changes involved, called chemical changes, are difficult to reverse. When wood burns, it undergoes a chemical change. It produces new substances, smoke and ash. However, it is very difficult to reverse the change. Compounds are substances that are made up of two or more elements whose atoms combine to form molecules. The molecules are broken down into atoms or changed into other molecules during a chemical change.

An element is a substance that is made up of a single kind of atom. These atoms cannot be broken down by chemical means. So elements are substances that cannot be separated into simpler substances by chemical changes.

◀ This figure of gold and copper alloy is about 20 cm (8 in.) high. It was made by the Indians of Colombia. Gold was one of the earliest known elements because it occurs naturally as nuggets. It is easily worked and has been used in jewelry for hundreds of years.

▼ The mineral galena is a compound of the elements lead and sulfur. Its molecules consist of an atom of lead and an atom of sulfur chemically bonded.

The Periodic Table

In the Periodic Table, elements are arranged in order of increasing atomic number. The atomic number is the number of protons in an element's nucleus. An element resembles those above and below it in the table – that is, those in the same "group." It also resembles the elements on each side of it – that is, those in the same "period." For example, helium, neon, argon, krypton, xenon, and radon are found in the group on the right-hand side of the table. All these gases are very similar and very unreactive. Moving to the left across the table, the elements gradually become more reactive. Those in the leftmost column are the most reactive of all.

Electron configurations

Scientists have found that the position of an element in the Periodic Table is related to the way the electrons are arranged.

The noble gases in the rightmost column of the table have eight electrons in their outer layer. This is a very stable arrangement, and it is difficult to remove electrons to allow chemical bonds, or links with other elements, to form.

The atoms of the elements in the leftmost column, such as sodium, have only a single electron in their outer layer. This electron can easily be removed, so the elements are very reactive. The atoms of the elements in the next-to-last group on the right, such as chlorine, have seven electrons in their outer layer. They can easily take up another electron, and are also very reactive. For this reason, sodium reacts violently with chlorine to form the chemical compound sodium chloride (common salt).

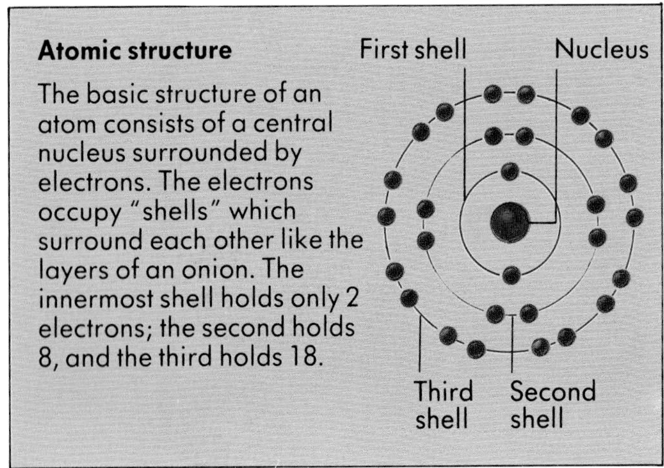

Atomic structure

The basic structure of an atom consists of a central nucleus surrounded by electrons. The electrons occupy "shells" which surround each other like the layers of an onion. The innermost shell holds only 2 electrons; the second holds 8, and the third holds 18.

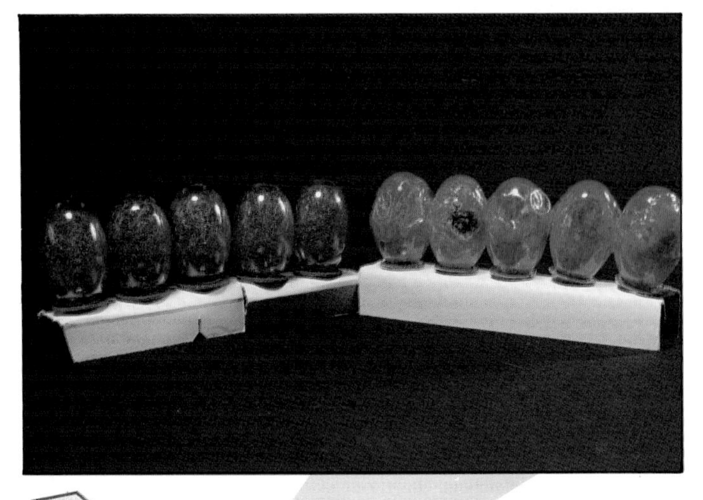

▲ Magnesium, when clean and freshly prepared, is a silvery metal found in seawater and several minerals. It burns with an intense white flame. Magnesium is used in fireworks, camera flashbulbs, and lightweight alloys.

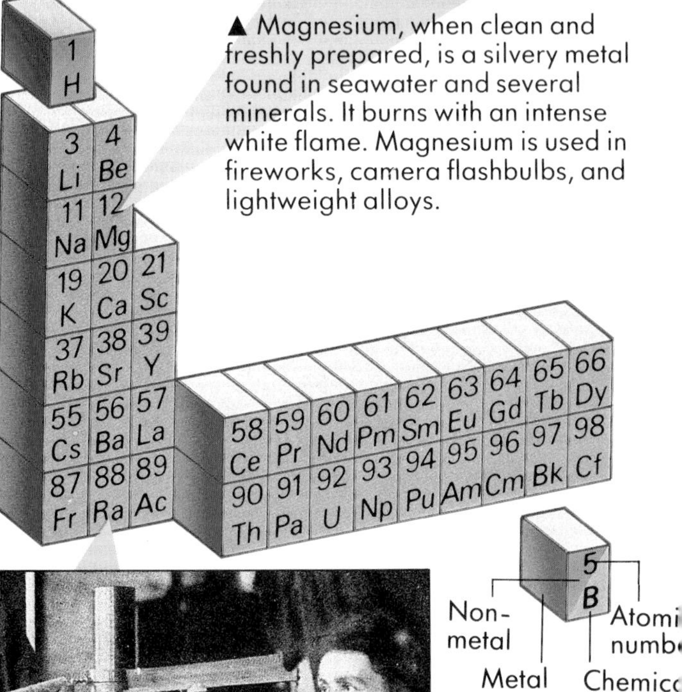

▲ Marie Curie and her husband Pierre discovered radium in 1899.

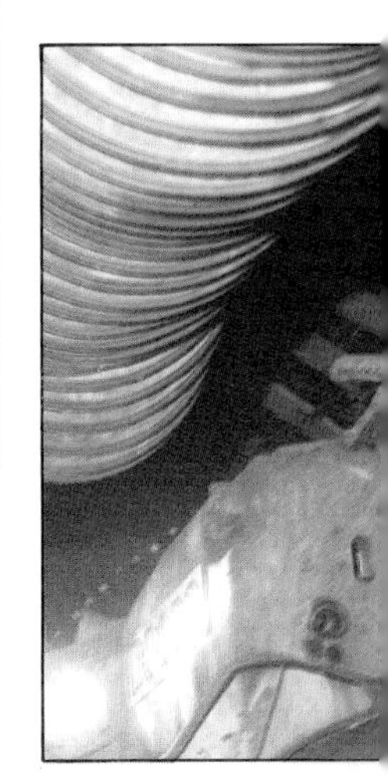

► Drilling tool, the tips of which are made of tungsten carbide. Tungsten is a very strong silvery-white metal.

◀ Parts of the Concorde airplane are made of titanium alloy. Titanium is a strong, light metal. It is found in many minerals.

▲ Gold is a very stable element, which does not corrode. For this reason, and because of its beauty and value, gold is used in jewelry.

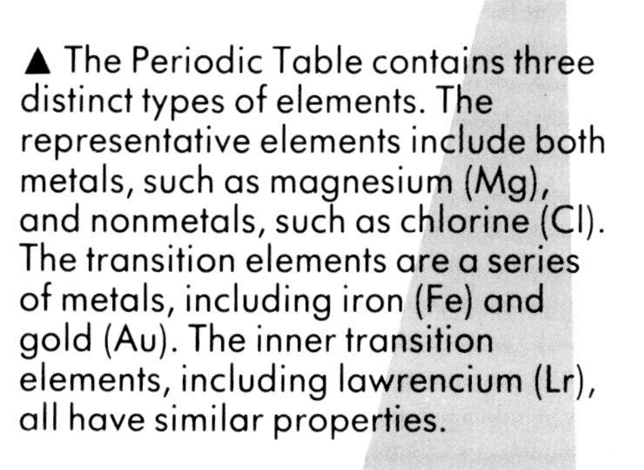

▲ The Periodic Table contains three distinct types of elements. The representative elements include both metals, such as magnesium (Mg), and nonmetals, such as chlorine (Cl). The transition elements are a series of metals, including iron (Fe) and gold (Au). The inner transition elements, including lawrencium (Lr), all have similar properties.

The elements

	Symbol	Atomic number		Symbol	Atomic number		Symbol	Atomic number
Actinium	Ac	89	Hafnium	Hf	72	Praseodymium	Pr	59
Aluminum	Al	13	Helium	He	2	Prometheum	Pm	61
Americium	Am	95	Holmium	Ho	67	Protactinium	Pa	91
Antimony	Sb	51	Hydrogen	H	1	Radium	Ra	88
Argon	Ar	18	Indium	In	49	Radon	Rn	86
Arsenic	As	33	Iodine	I	53	Rhenium	Re	75
Astatine	At	85	Iridium	Ir	77	Rhodium	Rh	45
Barium	Ba	56	Iron	Fe	26	Rubidium	Rb	37
Berkelium	Bk	97	Krypton	Kr	36	Ruthenium	Ru	44
Beryllium	Be	4	Lanthanum	La	57	Samarium	Sm	62
Bismuth	Bi	83	Lawrencium	Lr	103	Scandium	Sc	21
Boron	B	5	Lead	Pb	82	Selenium	Se	34
Bromine	Br	35	Lithium	Li	3	Silicon	Si	14
Cadmium	Cd	48	Lutetium	Lu	71	Silver	Ag	47
Calcium	Ca	20	Magnesium	Mg	12	Sodium	Na	11
Californium	Cf	98	Manganese	Mn	25	Strontium	Sr	38
Carbon	C	6	Mendelevium	Md	101	Sulfur	S	16
Cerium	Ce	58	Mercury	Hg	80	Tantalum	Ta	73
Cesium	Cs	55	Molybdenum	Mo	42	Technetium	Tc	43
Chlorine	Cl	17	Neodymium	Nd	60	Tellurium	Te	52
Chromium	Cr	24	Neon	Ne	10	Terbium	Tb	65
Cobalt	Co	27	Neptunium	Np	93	Thallium	Tl	81
Copper	Cu	29	Nickel	Ni	28	Thorium	Th	90
Curium	Cm	96	Niobium	Nb	41	Thulium	Tm	69
Dysprosium	Dy	66	Nitrogen	N	7	Tin	Sn	50
Einsteinium	Es	99	Nobelium	No	102	Titanium	Ti	22
Erbium	Er	68	Osmium	Os	76	Tungsten	W	74
Europium	Eu	63	Oxygen	O	8	Uranium	U	92
Fermium	Fm	100	Palladium	Pd	46	Vanadium	V	23
Fluorine	F	9	Phosphorus	P	15	Xenon	Xe	54
Francium	Fr	87	Platinum	Pt	78	Ytterbium	Yb	70
Gadolinium	Gd	64	Plutonium	Pu	94	Yttrium	Y	39
Gallium	Ga	31	Polonium	Po	84	Zinc	Zn	30
Germanium	Ge	32	Potassium	K	19	Zirconium	Zr	40
Gold	Au	79						

Forming crystals

Many substances are found in crystal form, with a regular arrangement of atoms. The metallic elements are examples. In metals, the atoms are packed closely together like stacks of identical balls. There are three ways that the atoms (balls) can be stacked together. For this reason, there are three different atomic arrangements found in metals. The first is called a body-centered cube. In this arrangement, the atoms are stacked in cube shapes, with an atom in the center of each cube. Sodium and vanadium crystals have this arrangement.

The second arrangement is called a face-centered cube. It is cube-shaped, with an additional atom in the center of each face. Aluminum and gold have crystals with this arrangement. The third is called hexagonal close-packed. It has the atoms arranged in hexagonal, or six-sided, groups. Magnesium has this arrangement.

The bonds, or links, between the atoms in a metal crystal are different from the bonds found in other crystals or molecules. The metal atoms are held together by a "sea" of electrons, which have broken free from the metal atoms. The electrons move freely among the atoms, acting as a kind of glue. The presence of free-moving electrons explains why metals are such good conductors of heat and electricity. The electrons carry the heat and electricity as they move about. Also, because the atoms are not strongly bonded to their neighbors, metals can be bent, hammered into sheets, and drawn into thin wire without breaking.

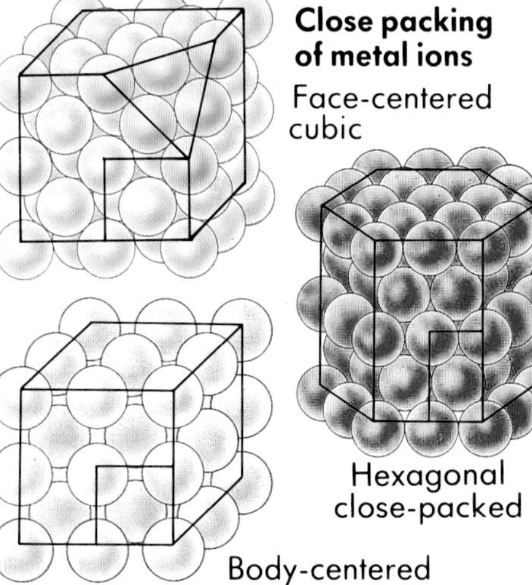

◄ Under a microscope, we can see the small crystals that make up a piece of aluminum.

▼ In an electron microscope, it is possible to locate the atoms (black dots) in the crystals.

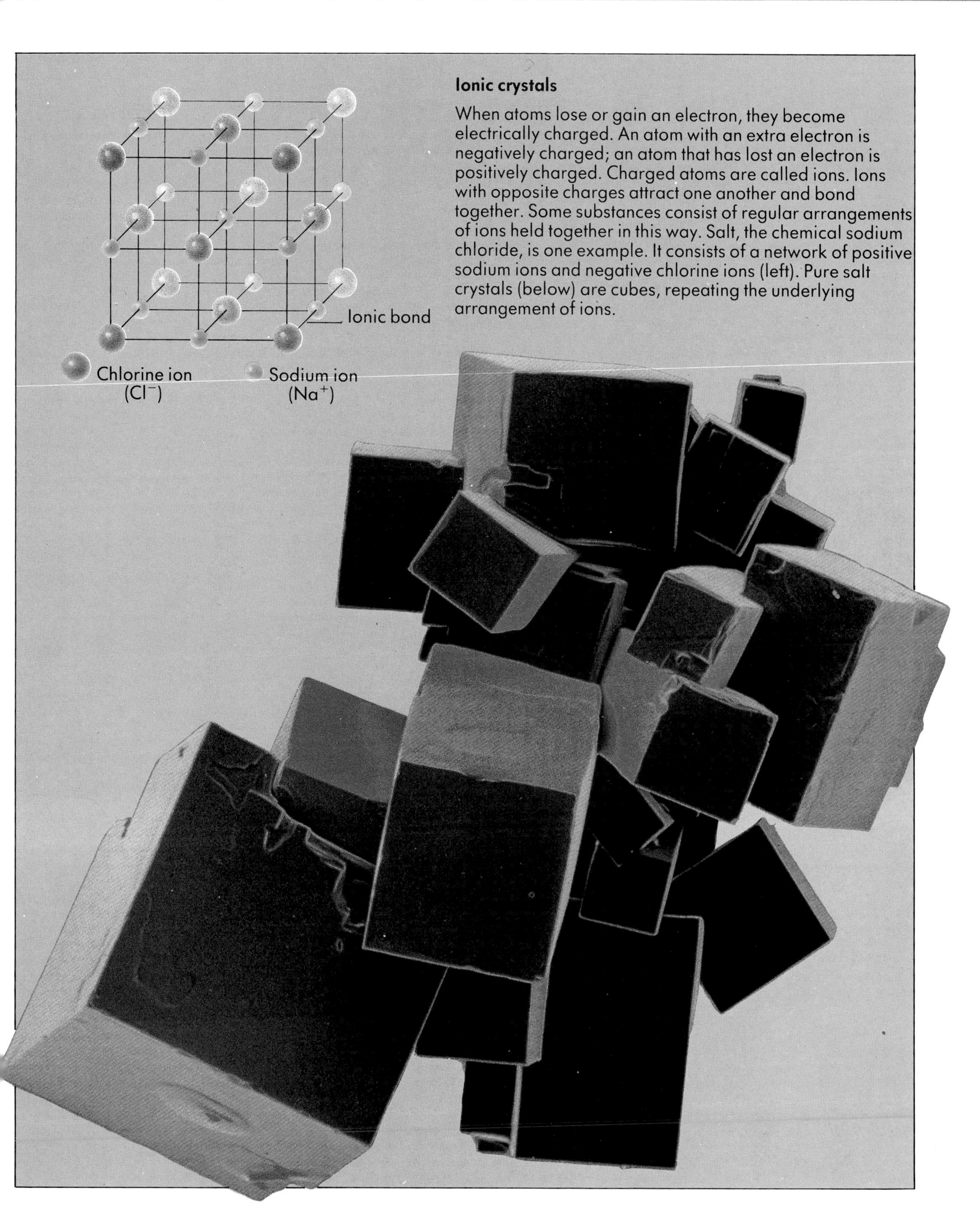

Ionic crystals

When atoms lose or gain an electron, they become electrically charged. An atom with an extra electron is negatively charged; an atom that has lost an electron is positively charged. Charged atoms are called ions. Ions with opposite charges attract one another and bond together. Some substances consist of regular arrangements of ions held together in this way. Salt, the chemical sodium chloride, is one example. It consists of a network of positive sodium ions and negative chlorine ions (left). Pure salt crystals (below) are cubes, repeating the underlying arrangement of ions.

Forming molecules

Atoms that readily gain or lose electrons to form ions do so in order to form a stable outer layer of electrons. An outer electron layer that has eight electrons is especially stable. For this reason, an atom with seven electrons in its outer layer, such as chlorine, can achieve a stable outer layer by gaining an electron. An atom with nine electrons in its outer layers, such as sodium, readily loses an electron in order to form a stable outer layer. The ions formed in this way readily group together to form ionic crystals. There are ionic bonds between the charged atoms in such crystals.

Covalent bonds

Chlorine atoms can also form another kind of bond in which electrons are shared between atoms. If two chlorine atoms come together, a pair of electrons, one from each atom, can be shared. This arrangement completes the outer electron layer of each atom and forms a stable molecule of two chlorine atoms. This type of bond, in which electrons are shared, is called a covalent bond.

Water molecules form because covalent bonds link two hydrogen atoms to an oxygen atom. The oxygen atom normally has six electrons in its outer layer, so it needs to share two electrons. One shared electron comes from each of the two hydrogen atoms. The hydrogen atoms form a stable electron layer with two electrons in it. It is unusual for atoms to be stable with two electrons in the outer layer. Only hydrogen and a few other elements are able to do this.

Scientists have a simple way of representing a water molecule. They write H_2O. This shows that the molecule is made up of two hydrogen atoms and one oxygen atom. But it does not show how the electrons are shared between the atoms. Other diagrams, using dots or short lines, are used to show how the electrons are shared and where the covalent bonds are. Even these diagrams do not show how the molecule really looks. Computer drawings are sometimes used to picture molecules more accurately.

Sometimes scientists make models of molecules, using balls to represent the atoms and sticks to show the bonds holding the molecule together. The most realistic models are called space-filling models. These use balls of different sizes to represent atoms of different sizes.

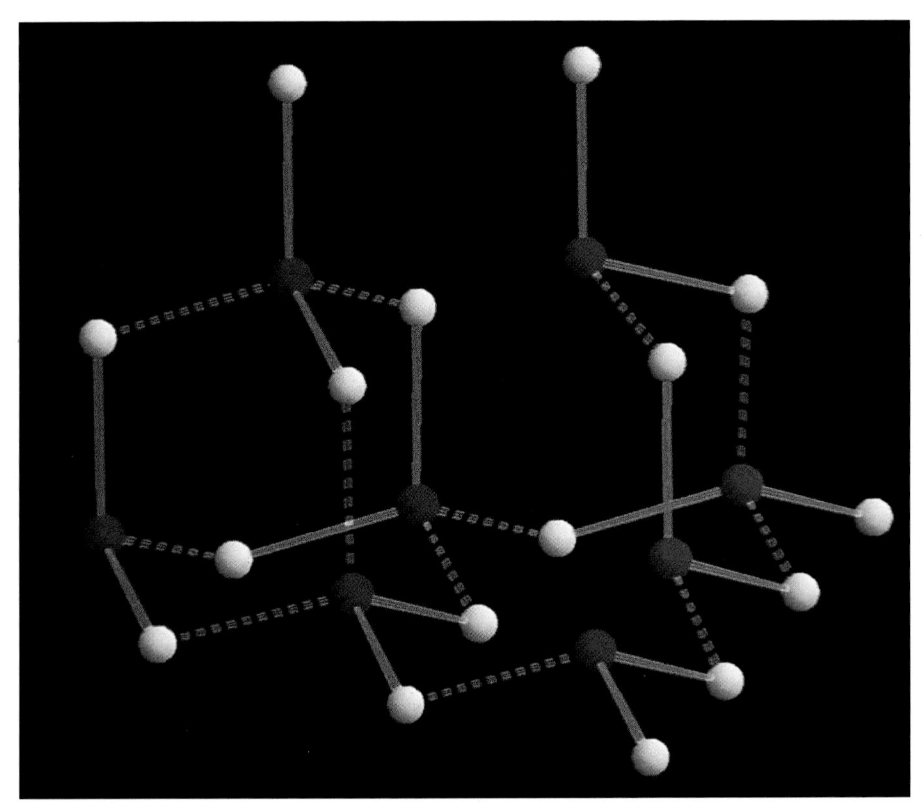

◀ ▼ In molecules of water, the oxygen atoms and hydrogen atoms are joined to each other by covalent bonds (shown in yellow). But separate molecules are also weakly bonded together. The hydrogen atoms in one molecule are weakly bonded to the oxygen atoms in other molecules by hydrogen bonds (shown in blue).

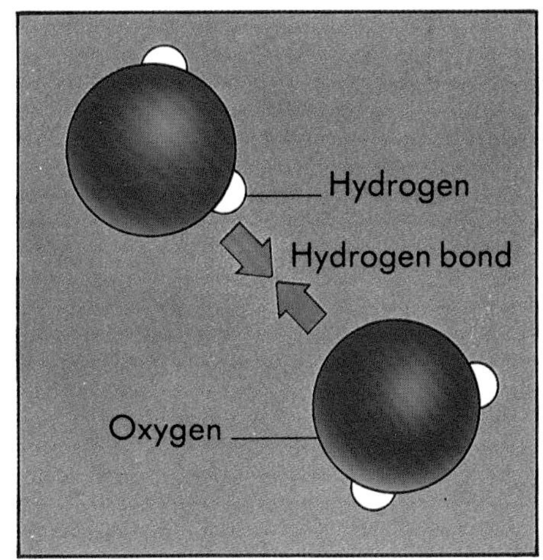

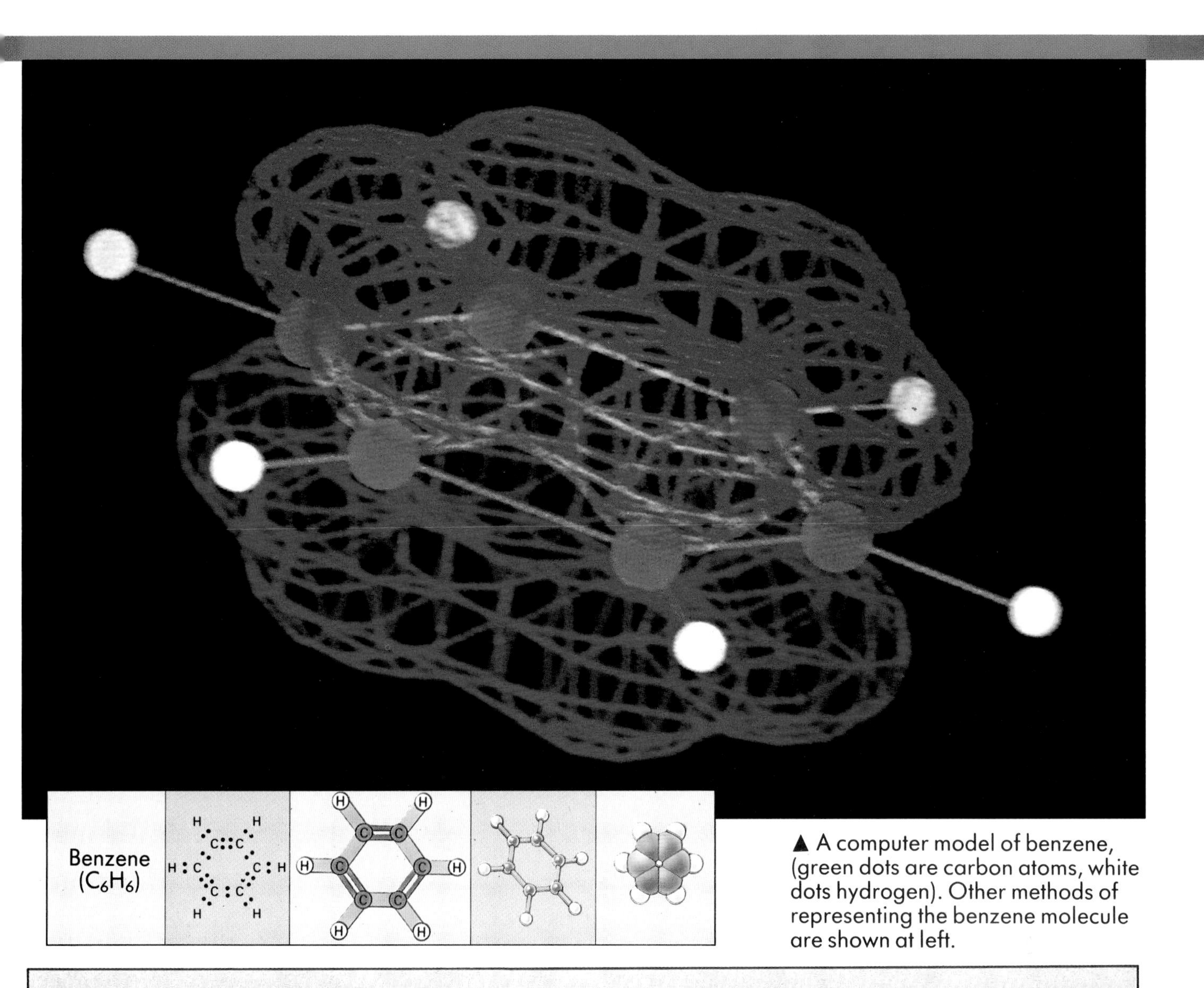

▲ A computer model of benzene, (green dots are carbon atoms, white dots hydrogen). Other methods of representing the benzene molecule are shown at left.

Types of chemical bond

There are three common types of chemical bonding, ionic, covalent, and metallic. Ionic bonding relies on the attraction of the opposite electrical charges of the ions. Covalent bonding uses shared electrons to link atoms in molecules, such as water. The linking of atoms in a metal crystal – metallic bonding – is brought about by a "sea" of free electrons moving around the atoms.

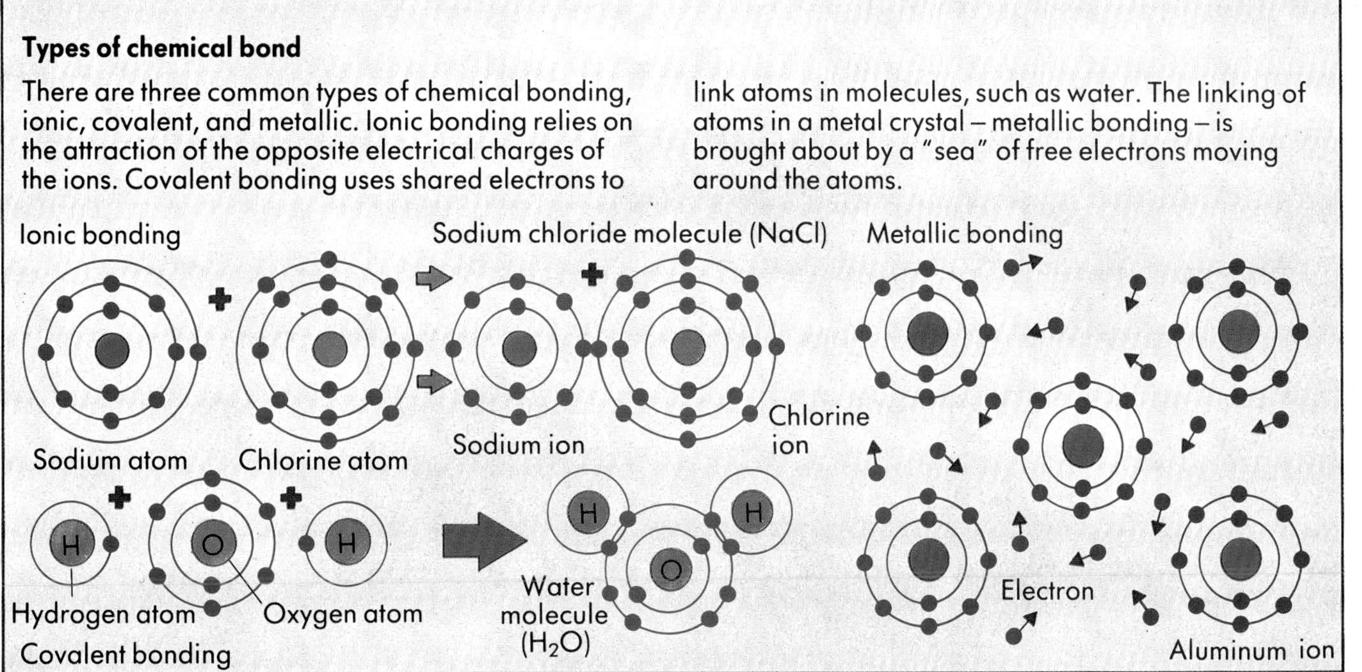

Chemical reactions

When coal burns, or dynamite explodes, or a metal ship rusts, a chemical reaction, or change, takes place. These changes involve the formation or breaking up of molecules. When coal burns, for example, carbon atoms in the coal link up with oxygen atoms from the air to form molecules of carbon dioxide. Scientists use a kind of chemical shorthand to write an equation to show the process:

$$C + O_2 \rightarrow CO_2$$

Coal is made up of carbon. The chemical symbol for an atom of carbon is C. Air contains oxygen molecules made up of two oxygen atoms. These molecules are written as O_2. The carbon dioxide molecule that is formed is represented as CO_2. It is made up of one carbon atom joined to two oxygen atoms.

When the gas methane burns in air, it produces carbon dioxide and water. The reaction describing this process is:

$$CH_4 + 2O_2 \rightarrow CO_2 + 2H_2O$$

This equation is more complicated. The methane molecule is written as CH_4. It is made up of a carbon atom joined to four hydrogen atoms. The equation shows how the methane molecule reacts with two molecules of oxygen to produce a carbon dioxide molecule and two molecules of water represented by $2H_2O$.

Interacting molecules

All chemical changes involve energy changes. Many reactions need a supply of heat energy before they can start. Coal must be heated before it begins to burn, for example. In general, once a reaction starts, it may produce heat, or it may continue to take in heat and require heat to make any change take place.

In a molecule, the electrons are arranged in layers, or orbitals, as in an atom. But the shapes of the orbitals are more complicated because there is more than one atom in a molecule. In some molecules, the electrons are arranged so that they have lots of energy. These are high-energy molecules, such as methane.

If the atoms in a high-energy molecule are rearranged, the energy can be released. One way of doing this is to break the molecule into parts and reassemble the parts into low-energy molecules. This is what happens when methane, for example, is burned. In plants, a series of chemical reactions converts low-energy carbon dioxide gas from the air into high-energy sugar molecules, which the plant uses for food. This process, which takes place only in sunlight, is called photosynthesis.

▼ All chemical reactions involve energy changes. Some reactions, such as the burning of waste methane gas in the desert near an oil field, give off energy. Other reactions extract energy from their surroundings. Photosynthesis extracts energy in the form of sunlight.

Speeding up chemical reactions

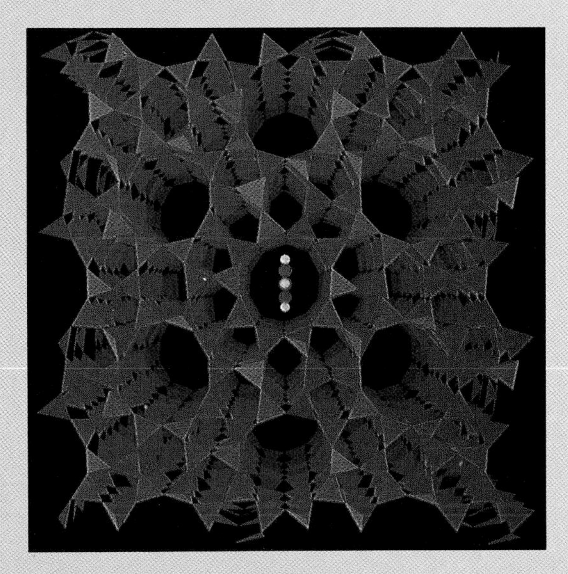

A catalyst is a substance that alters the speed of a chemical reaction without itself being used up.

Minerals known as zeolites are used in New Zealand as catalysts in the manufacture of gasoline from the alcohol methanol. Inside the zeolite (above) are minute channels and cavities. The methanol molecules enter these channels, where they give up their oxygen atoms, combine, and form gasoline. Zeolites are also often used in water softening They change hard water to soft by taking out certain chemicals.

◀ The demolition of a hotel. Explosives release energy much faster than it can spread into the surroundings. This raises the temperature, increasing the speed of the reaction even further. The reaction creates a large volume of expanding gas, and an explosion takes place.

Molecules in motion

Spot facts

- *There is only one temperature at which molecules stop moving: −273.15°C (−459.7°F). This is the lowest temperature of all, called absolute zero.*
- *At very low temperatures, around −270°C (−455°F), the gas helium becomes a liquid. A cupful of liquid helium empties as the liquid flows up the side of the cup. It is like water flowing uphill.*
- *There is more heat energy locked up in an iceberg than in a cup of boiling water. This is because the iceberg, although it is colder, is much larger.*
- *Sounds travel through air at a speed of about 1,224 km/h (761 mph), depending on the temperature. Sound travels 15 times faster through steel than it does through air.*

The molecules in objects around us are always moving. This movement is a form of energy that we call heat. The molecular movements are random vibrations – small back-and-forth movements. The higher the temperature, the faster the molecular vibrations. But molecules can also move in more regular ways. Sounds are caused by molecules of the air vibrating in a regular, wavelike manner. We can use sound waves to carry information, as in speech. We also get pleasure from music because of the regular vibrations produced by musical instruments. Sound is important to us as a way of communicating. Animals, too, use sound to communicate. In addition, some animals use very high-pitched sounds to find their way around. Such ultrasounds are also finding many uses in science and medicine.

► Drums and cymbals are percussion instruments. They produce sounds when tightly stretched skin or metal is struck. The vibrating metal or skin makes molecules of air vibrate. The vibrating air molecules create the sounds.

Heat and temperature

There is a difference between the temperature of an object and the amount of heat it contains. The temperature measures how fast the molecules of the object are moving. In a hot object, the molecules vibrate back and forth at great speed. In a cold object, they vibrate more slowly. But the amount of heat energy in an object is the total energy of all its molecules. This depends upon both the temperature and the amount of material in the object. Even though its temperature is higher, a very hot small object does not contain as much heat as a larger object that is cooler. A large iceberg contains enough heat to boil water if only it could be collected and used.

There are three common scales for measuring temperature. On the Celsius, or centigrade, scale, the temperature at which water freezes is 0°C, and water boils at 100°C. The scale is named for the Swedish physicist Anders Celsius. On the Fahrenheit scale, named after the German scientist Gabriel Fahrenheit, the freezing point of water is 32°F, and the temperature of boiling water is 212°F. The third scale is called the Kelvin scale, for a Scottish scientist, Sir William Thomson, who became Lord Kelvin. On this scale, the lowest possible temperature, called absolute zero, is 0K. One kelvin is equal to 1°C. On the Kelvin scale, 0°C becomes 273K, and 100°C is 373K.

▼ In a mercury or alcohol thermometer, the liquid in the bulb expands and moves up the thin tube. The scale indicates the temperature. In a bimetallic thermometer, a strip made of steel and brass bonded together is used. The brass expands more than the steel as the strip gets warmer. This makes the strip bend, moving the needle over the scale. The thermocouple thermometer is made from wires of two different metals. If the welded ends of the wires are at different temperatures, a small voltage is produced. This moves the needle over the meter scale. The digital thermometer has a thermocouple in its probe, and an integrated circuit creates a digital readout.

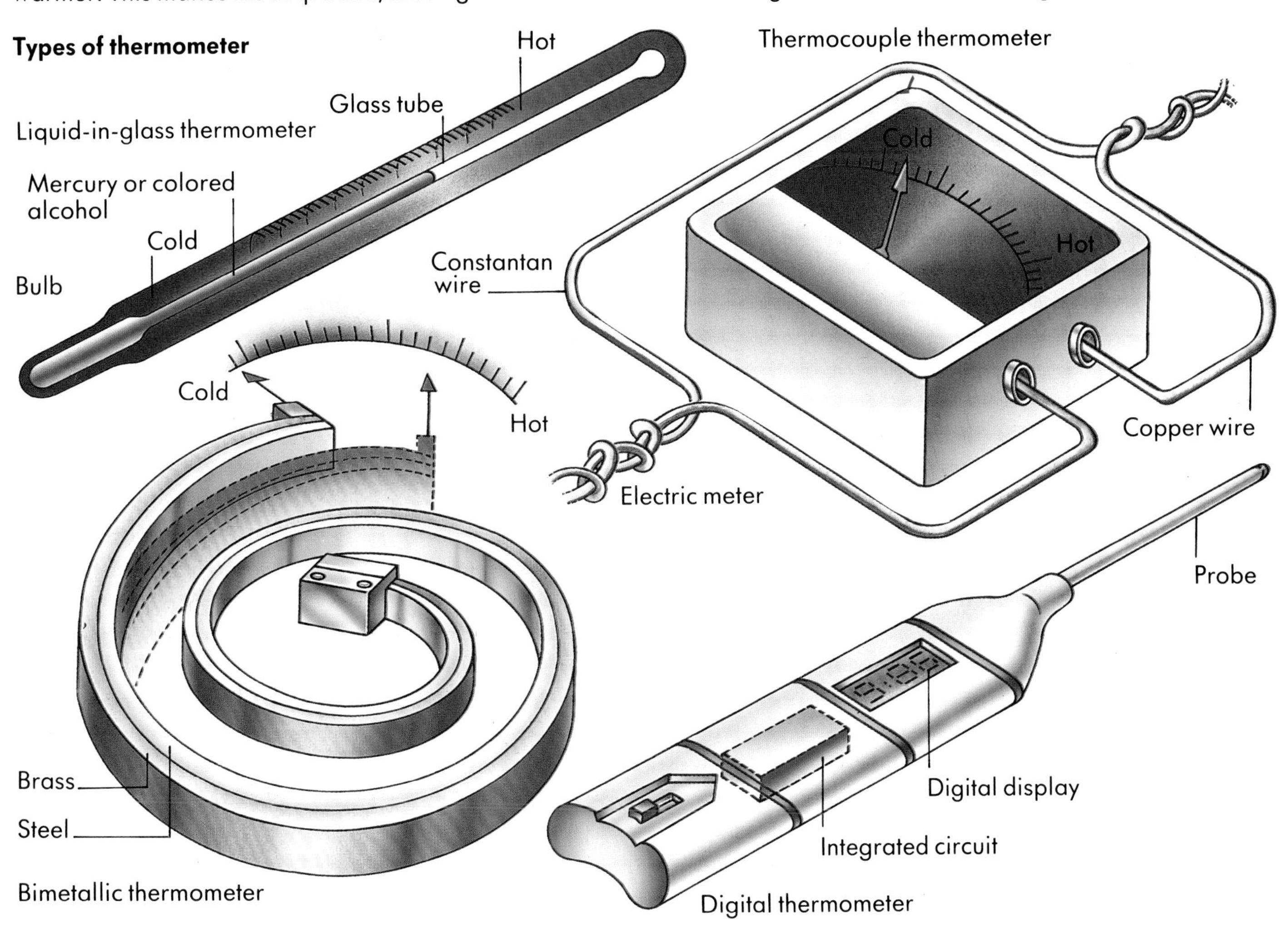

Travelling heat

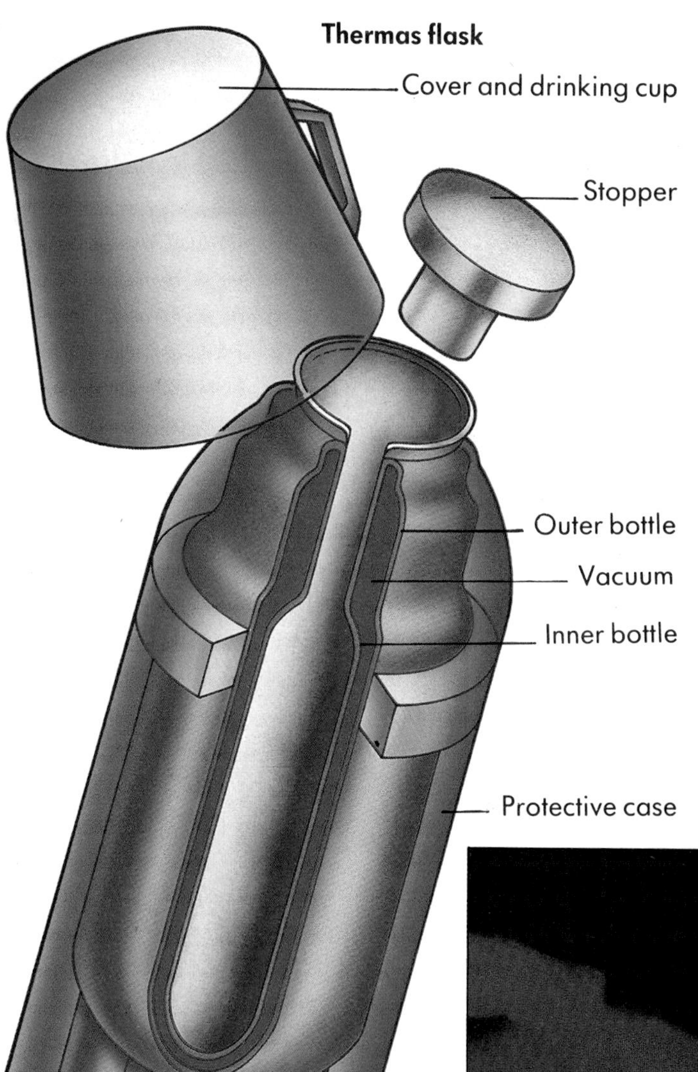

Heat can travel in three ways: by convection, conduction, and radiation. Convection and conduction can take place only where there is matter. They are processes that involve moving molecules. But heat can also travel through empty space, where there is no matter.

Heat travels through empty space as energy-carrying waves, called radiation or infrared rays. William Herschel, a famous English astronomer, discovered infrared rays about 1800. Using a prism, he split the light coming from the Sun into the colors of the rainbow. He noticed that heat was coming from the red light. He realized that there must be invisible rays carrying heat in the sunlight. He called them infrared rays.

Infrared rays are very similar to light rays and travel at the same speed, 300,000 km/s (186,000 mps). Like light rays, they can be reflected and absorbed by matter. Infrared rays are given off by all objects. They are given off by our bodies, for example, and doctors can use the rays to detect some diseases.

▲ The thermos, or vacuum bottle, keeps liquids hot or cold because its walls are airless. Heat that would normally pass from or to the contents of the flask by convection or conduction is prevented from doing so by the vacuum.

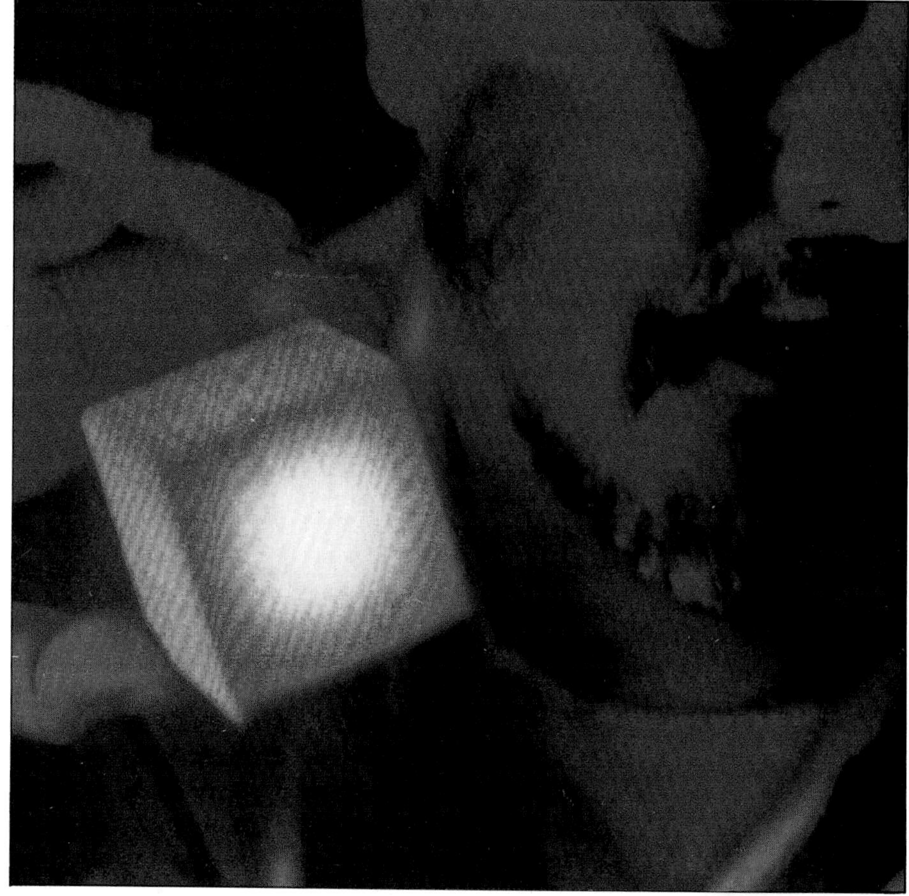

► The NASA space shuttle is covered with tiles made of silica, a poor conductor of heat. The tiles protect the craft from the high temperatures produced when it reenters the Earth's atmosphere.

Convection carries heat in liquids and gases. When liquids and gases are heated, they expand and become less dense. The warmer material rises. This movement of heated material sets up a current, called a convection current. Such currents are seen on a large scale in the atmosphere, where they cause winds and breezes. Convection currents in the oceans carry heat from the tropics to colder parts of the world. On a smaller scale, convection currents spread heat through liquid in a saucepan.

When a solid object is heated, molecules near the source of heat vibrate more rapidly than those farther from it. The rapidly vibrating molecules bump into the molecules next to them, passing on some of their energy. These molecules, in turn, bump against their neighbors. In this way the heat flows to all parts of the object. This heat-spreading process is called conduction. Some materials, such as metals, are good conductors of heat. Materials which do not conduct heat easily, such as rubber, are called insulators.

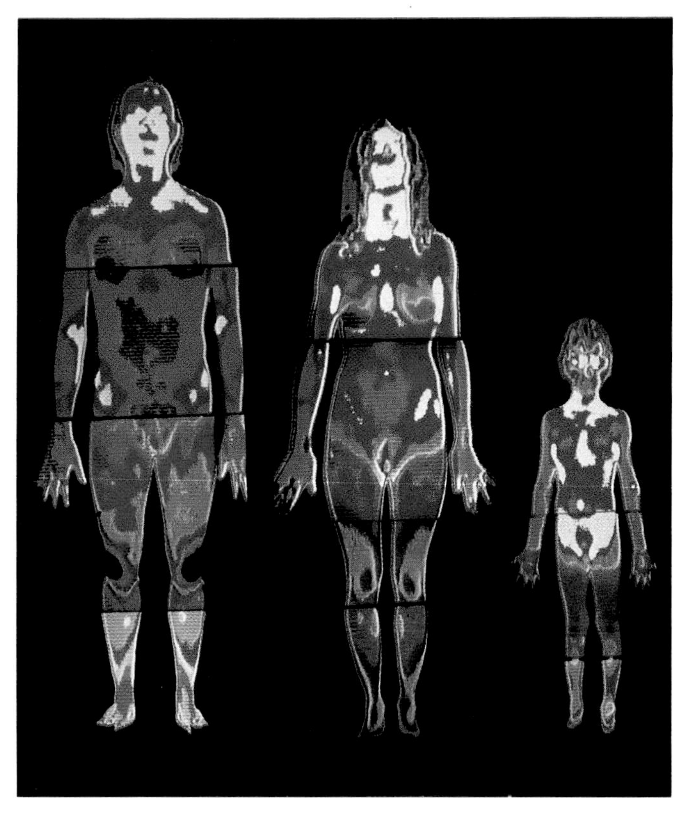

▲ A thermal image of a family. Hot objects give off infrared radiation, which can be photographed. The hottest parts of the people's bodies are brightest.

▼ A glider pilot must seek out rising currents of warm air, called thermals, to carry the glider upward. Gliding birds also use these convection currents to gain height.

Sound

All sounds are made by vibrating objects. For example, when you speak, your vocal chords vibrate. As a vibrating object moves, it pushes the air molecules in front of it. This creates a region of high pressure, where the air molecules are pressed together. When the object moves backward, it creates a region of low pressure, where the molecules are farther apart than normal. These high- and low-pressure regions travel out from the vibrating object as sound waves, like ripples on a pond.

When the sound waves enter a person's ear, an intricate system of tiny bones and a very thin membrane causes the eardrum to vibrate, and the sound is heard. The size of a vibration or wave is called its amplitude. The greater the amplitude of a sound wave, the louder it is. The loudness of a sound is measured in units that are called decibels.

Frequency

The number of complete vibrations or waves per second in a sound is called its frequency. The greater the frequency of a sound wave, the higher the pitch of the sound we hear. Frequency is measured in units called hertz, for Heinrich Hertz, who worked on radio waves between 1885 and 1889. One hertz equals one complete vibration per second.

Most people can hear sounds with frequencies as low as 20 hertz. Young people can hear sounds with frequencies up to 20,000 hertz, but most older people can hear sounds only up to 16,000 hertz. Cats can hear sounds up to 25,000 hertz and dogs do even better – they can hear sounds up to 35,000 hertz.

Sounds travel faster through liquids and solids than through gases. In seawater, for example, the speed of sound is 1,500 m/s (4,900

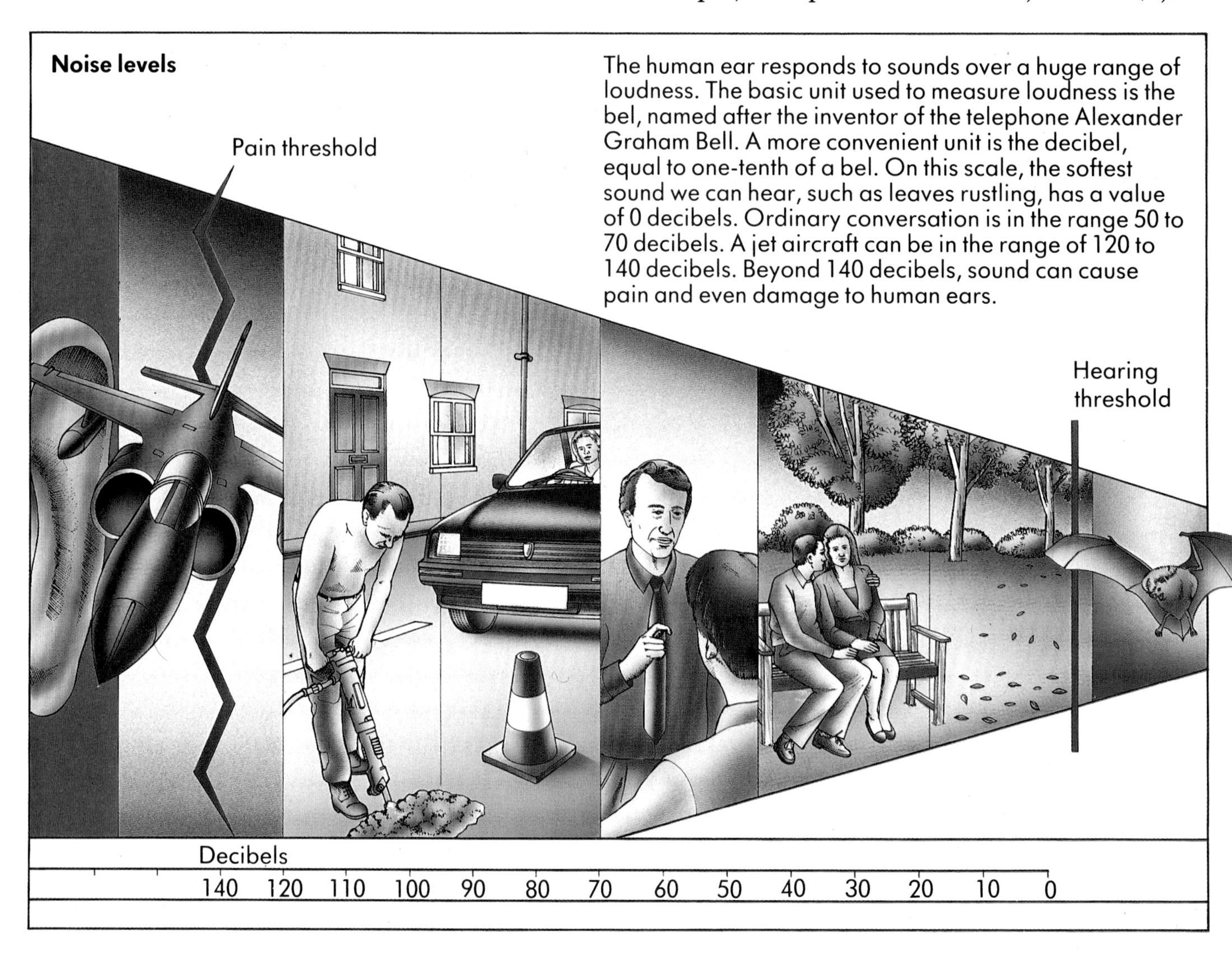

Noise levels

The human ear responds to sounds over a huge range of loudness. The basic unit used to measure loudness is the bel, named after the inventor of the telephone Alexander Graham Bell. A more convenient unit is the decibel, equal to one-tenth of a bel. On this scale, the softest sound we can hear, such as leaves rustling, has a value of 0 decibels. Ordinary conversation is in the range 50 to 70 decibels. A jet aircraft can be in the range of 120 to 140 decibels. Beyond 140 decibels, sound can cause pain and even damage to human ears.

fps), over four times the speed in air. The speed of sound in air is a little less than 350 m/s (1,100 fps). Speed also depends upon temperature. The higher the temperature, the greater the speed. This is why sounds seem louder and travel farther at night than during the day. At night, the air near the ground is cooler than the air above it. Sound waves traveling upward into the warmer air are bent back toward the ground, carrying sound a greater distance. The bending of waves in this way is called refraction.

Sound waves can also be reflected. They can bounce off surfaces they hit, causing echoes. Sounds, like all waves, can also bend around corners. This is called diffraction. The amount a wave is bent depends upon its frequency. Lower frequencies are bent more than higher ones. This is why words overheard round the corner of an open door sound mumbled.

▼ When the string of a piano is struck, the profile of the string assumes a wave pattern. This is called a standing wave. The simplest standing wave is the fundamental. The higher-pitched harmonics correspond to standing waves of higher frequency.

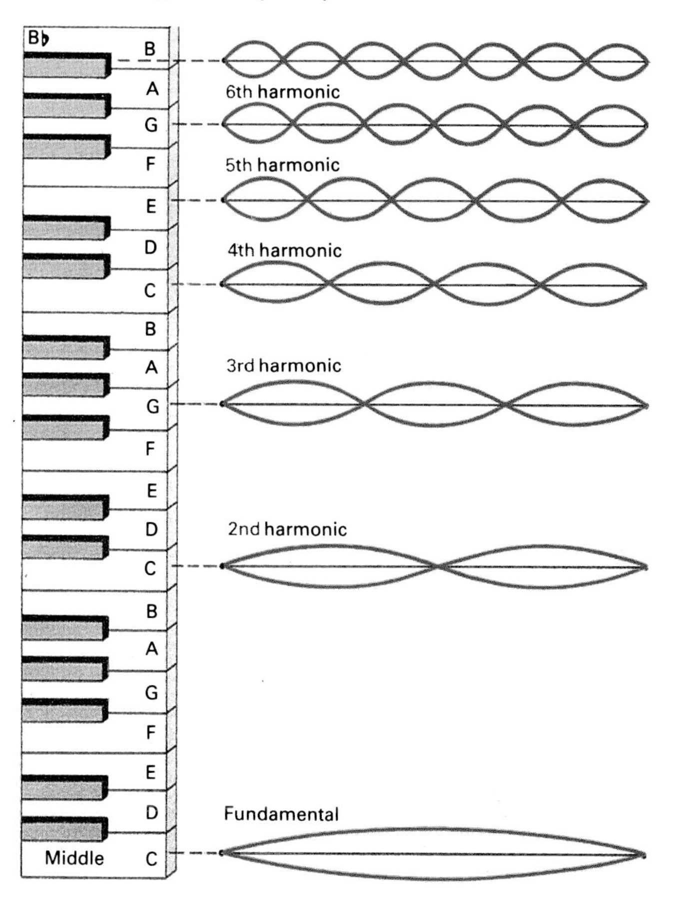

The Doppler effect

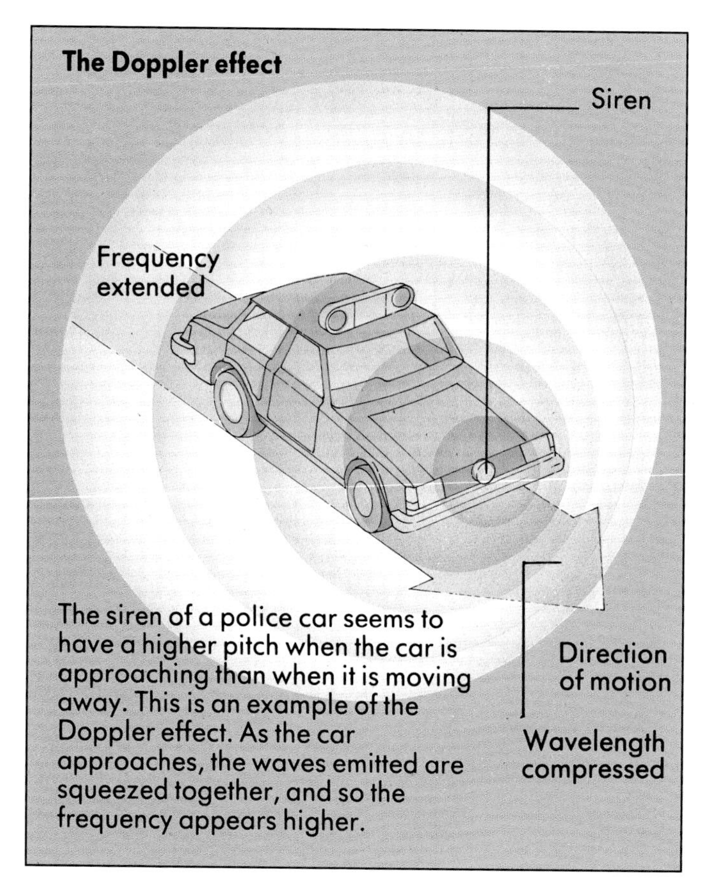

The siren of a police car seems to have a higher pitch when the car is approaching than when it is moving away. This is an example of the Doppler effect. As the car approaches, the waves emitted are squeezed together, and so the frequency appears higher.

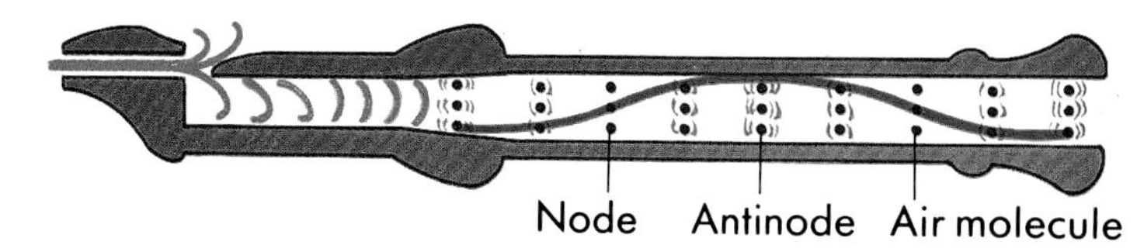

The recorder

▲ When somebody plays a wind instrument, such as a recorder, a standing wave is produced inside the instrument. It consists of a stationary pattern of vibrating air. A point where there is no vibration is called a node; an antinode is where vibration is greatest.

Silent sound

Some sound waves are silent – we cannot hear them. Their frequencies fall outside the range the human ear can detect. These waves are called infrasonic and ultrasonic. Infrasonic waves have frequencies below 20 hertz. Infrasounds are produced by tremors in the Earth's crust, by waterfalls and seaside waves, and by wind passing over mountain ranges. These sounds can carry for hundreds of kilometers. Birds can detect them and may use them to navigate. Elephants and whales use infrasound to communicate over long distances.

Ultrasonic waves have frequencies above 20,000 hertz. Sounds with these high frequencies are common in nature. Rats, mice, and shrews make ultrasonic squeaks with frequencies up to 100,000 hertz. Such sounds cannot carry far and are ideal for rodents, allowing them to communicate in their burrows without being overheard by their enemies outside.

The oilbird of South America, which spends much of its time in caves, makes high-frequency clicking sounds and can locate obstacles such as the walls of its cave by listening to the echoes.

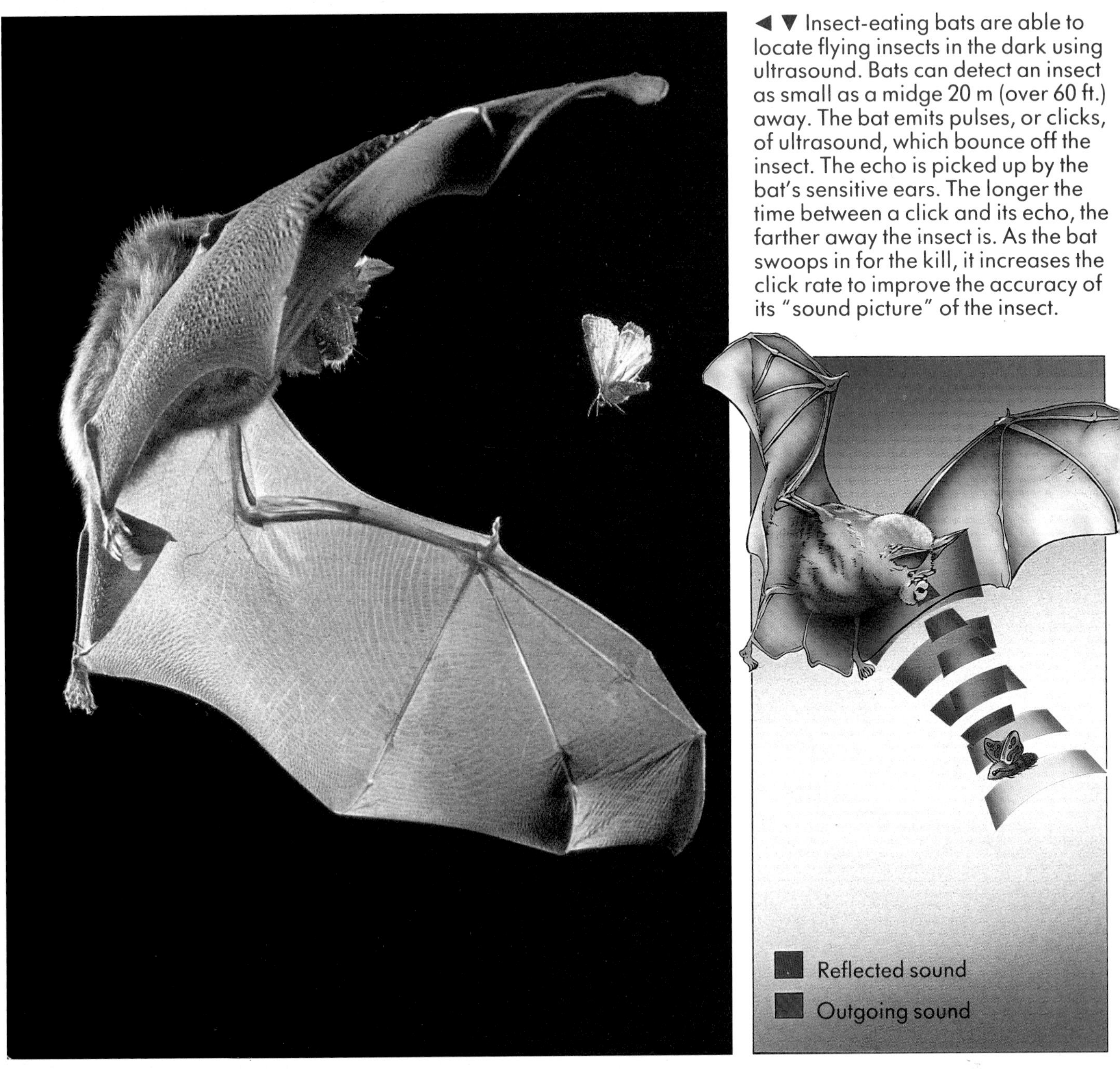

◀ ▼ Insect-eating bats are able to locate flying insects in the dark using ultrasound. Bats can detect an insect as small as a midge 20 m (over 60 ft.) away. The bat emits pulses, or clicks, of ultrasound, which bounce off the insect. The echo is picked up by the bat's sensitive ears. The longer the time between a click and its echo, the farther away the insect is. As the bat swoops in for the kill, it increases the click rate to improve the accuracy of its "sound picture" of the insect.

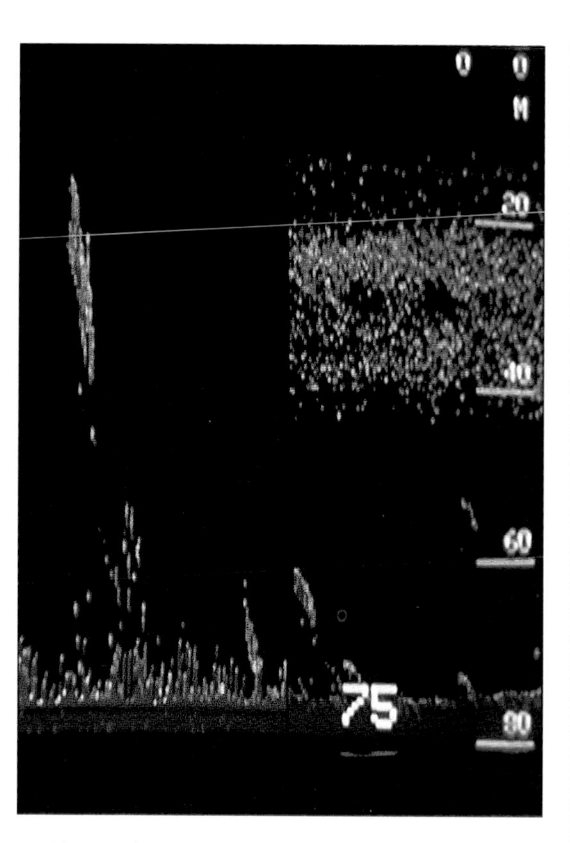

▲ The undersea picture of shoals of fish was produced by a high-detail echo-sounding system. The image features deepening color, from blue through yellow to red, to denote increasing fish numbers.

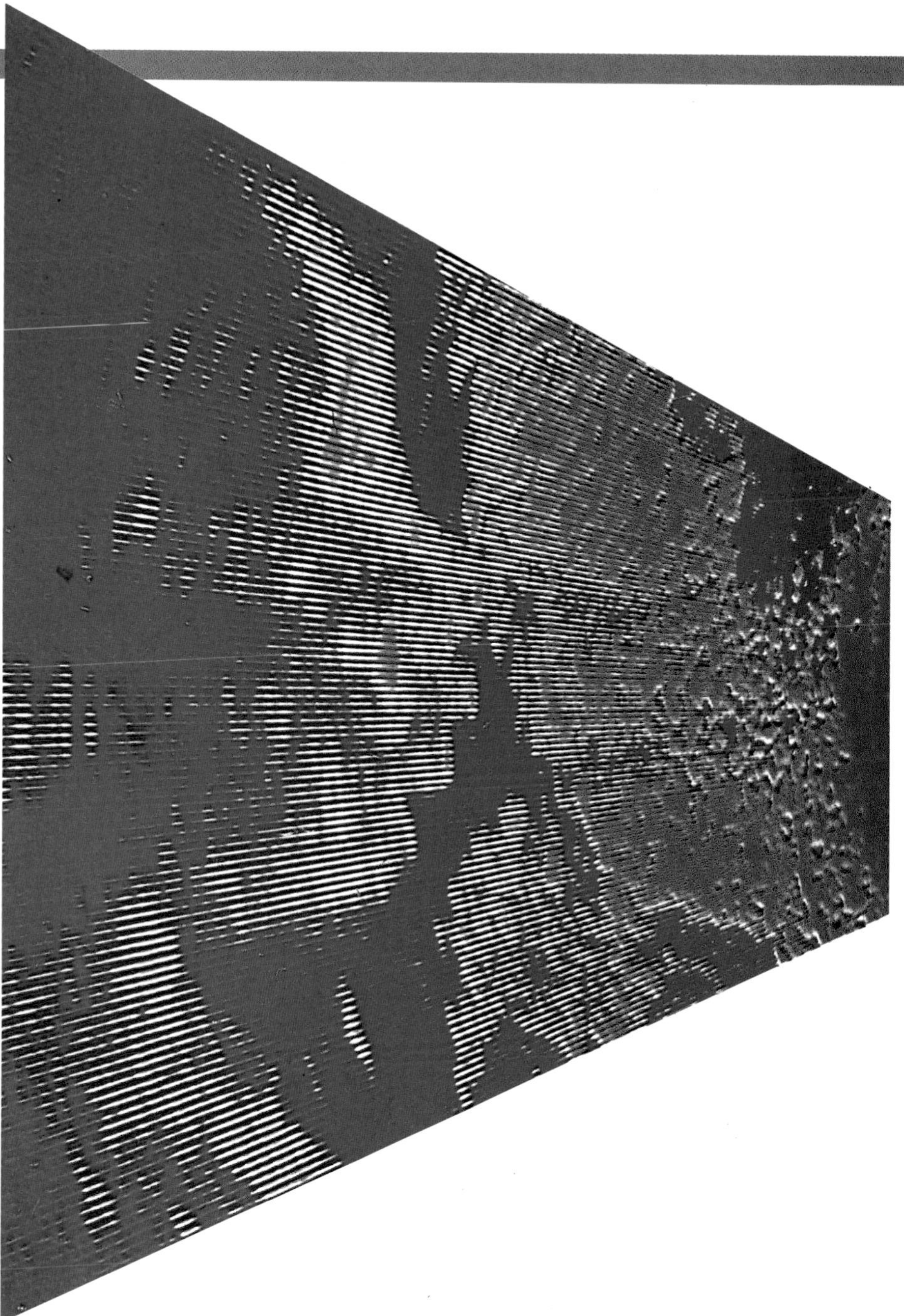

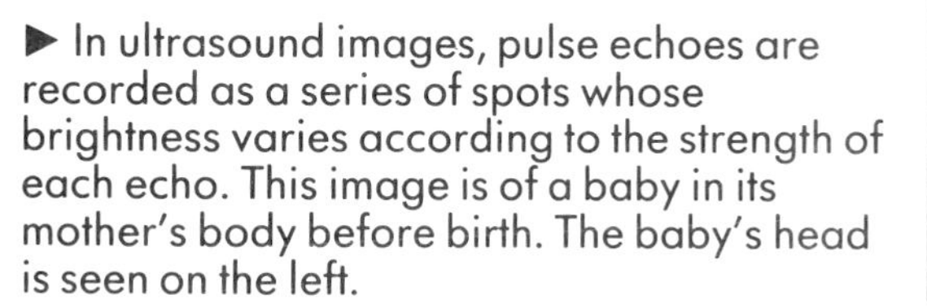

► In ultrasound images, pulse echoes are recorded as a series of spots whose brightness varies according to the strength of each echo. This image is of a baby in its mother's body before birth. The baby's head is seen on the left.

Ultrasound and its uses

There are many ways that we can use ultrasound. One of the earliest was in an echo sounder, used to measure the depth of the sea under a ship. In an echo sounder, a beam of ultrasound is directed down at the seabed. By timing the echo, the depth of the sea can be calculated. Shoals of fish, and submarines, can be located in the same way. In industry, ultrasound is used to measure the thickness of metal sheets and to detect flaws in metals.

In ultrasonic welding, a beam of ultrasound is focused onto a metal joint. Sufficient heat is generated at the joint to melt the metal. Holes of any shape can be bored in metal using ultrasonic drills. There is even a microscope that uses ultrasound instead of light.

Ultrasound is used in medicine, too. Pregnant women are often given ultrasound scans to see if their baby is developing normally. A beam of ultrasound is swept across the woman's body, and the reflected signal is picked up and shown on a television-type screen. This technique can also be used to detect disease in organs such as the kidney, and to examine the brain for damaged blood vessels. In many ways, ultrasound scanning is like a safe X-ray examination. However, it cannot detect very fine detail.

Part Two

The universal forces

Electricity provides the most useful form of energy there is. It holds the key to our modern civilization. More fundamentally, it also holds the key to the makeup of matter. The atoms of matter are made up of electrically charged particles, and are largely held together by electrical attraction. Electricity is thus a great universal force, dictating the nature of matter everywhere.

Closely linked to electricity is magnetism, another great force in the Universe. The two go hand in hand, and together are the subject of the science of electromagnetism. They travel together through space as a wave motion, forming a family of electromagnetic waves. Most familiar are the waves by which we see – light.

Another great force acts throughout the whole Universe and in effect holds the Universe together. It is called gravity.

◀ Patterns created when scratched metal foil is flattened between glass. These patterns are called interference patterns because they occur when light waves interfere with each other.

Electricity

Spot facts

- *About 3 quintillion electrons pass through a burning light bulb every second.*
- *Electric eels store enough electricity in their tails to light up 12 light bulbs.*
- *A household light bulb would have to shine for 10,000 years to release the same amount of light energy as a flash of lightning.*
- *Electrons are not only particles, they are also waves! Electron waves are used instead of light waves in electron microscopes.*

The world is built from atoms made from a small number of different particles. Most of the particles carry electric charges. It is the force between these charges that helps to hold atoms together. Taming the energy held in these charges – seen uncontrolled in a lightning flash – has been one of the most important successes of science. Electricity has become our obedient servant, lighting our cities, turning the wheels of industry, and powering our household gadgets. New ways of controlling electricity, using materials called semiconductors, have made possible modern electronics, as seen in radios, record players, televisions, and computers.

► The bright lights of Los Angeles. Electricity is the servant and messenger of the modern world. It is the most convenient source of power ever discovered. It supplies light and heat as well as mechanical power. It is also easily and inexpensively carried along wires or cables.

Electric charges and fields

The origin of electricity lies inside the atoms that make up matter. Electrons and protons carry tiny amounts of electricity. They are said to have an electric "charge." There are two kinds of electric charge. The electron has one kind, called a negative charge. Protons have the other kind, called a positive charge. If two negative charges or two positive charges are brought close together, they repel each other and push apart. If a positive charge is brought near a negative charge, the charges attract each other and pull together. In other words, like charges repel each other, and unlike charges attract each other.

Normal atoms have no overall electric charge. The same number of negatively charged electrons orbit the nucleus as there are positively charged protons in it: the negative and positive charges are balanced. However, atoms and the objects that they make up can become electrically charged by losing or gaining electrons. If an object gains electrons, it becomes negatively charged. If the object loses electrons, it becomes positively charged.

Electric fields

The region around a charged object or particle, where the electric force can be determined, is called an electric field. The strength of the field at a particular point depends on the size of the charge on the object and the distance between the point and the object. The field becomes stronger the closer the point is to the object.

Lines of force

An electric field can be represented by lines of force in space. Arrows on the lines show the direction of the force. The lines run out of positive charges (2) and into negative charges (1). The lines show how like charges repel (3), whereas unlike charges attract (4). The lines between flat plates are parallel (5).

1

2

3

4

5

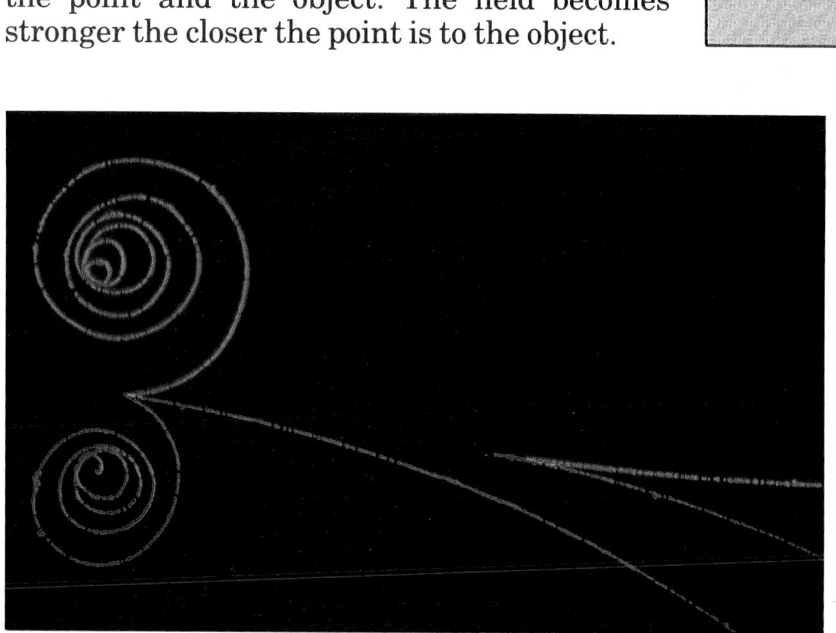

◀ Here, particles with opposite charges move in a bubble chamber. A magnetic field makes the particle tracks spiral in different directions. The green tracks are made by electrons with a negative charge. The red tracks are made by positrons, particles with the same mass as electrons but carrying a positive charge. Electrons are normally found orbiting the central nucleus of an atom. Electrons are also the origin of electricity.

Static electricity

If you rub a balloon on a woolen sweater, you give the balloon an electric charge. The rubbing transfers electrons from the sweater to the balloon. The balloon gets a negative electric charge because of its extra electrons. If you hold the balloon up to a wall, the balloon sticks to it. The negatively charged balloon is attracted to the positive charges in the wall. If you rub two balloons on wool and put the balloons next to each other, they push apart. This shows that similar charges repel each other.

In these experiments, you have been making and using static electricity – electricity that does not move but stays in one place. The ancient Greeks, about 2,500 years ago, did similar experiments by rubbing a piece of amber – a fossilized resin material – with fur. The word "electricity" comes from *elektron*, the Greek word for amber.

When you walk across some types of nylon carpet, static electricity builds up on you. If you touch something metallic, small sparks will jump from you to the metal. When you take off a nylon shirt or blouse, you can sometimes hear a crackling sound, and see small sparks. These sparks are like miniature lightning flashes.

Lightning conductor

In 1752 the famous American scientist and inventor Benjamin Franklin did a dangerous experiment. He flew a kite during a lightning storm. Electricity flowed down the string of the kite, making a small spark on a metal key near his hand. This showed that lightning was just a large electric spark. Later, Franklin used his discovery to invent the lightning rod. This is a metal rod at the top of a tall building that is connected by a cable to the ground. It carries the electricity safely away if lightning should strike the building.

► Lightning is caused when a large electric charge builds up on a cloud as the result of ice and water particles in the cloud rubbing together. Positive charges build up at the top of the cloud and negative electrons at the bottom. The electrons suddenly leap from the cloud to the ground, or to another cloud.

▲ An electronic flashgun uses devices called capacitors to store electric charges. Capacitors are also used in computer memories and radio circuits.

◄ This child's hair is standing on end because it is electrically charged. Some electrons have rubbed off her hair onto the comb, giving her hair a positive charge. Because each of her hairs has the same charge, they repel each other and stick out.

Electrons on the move

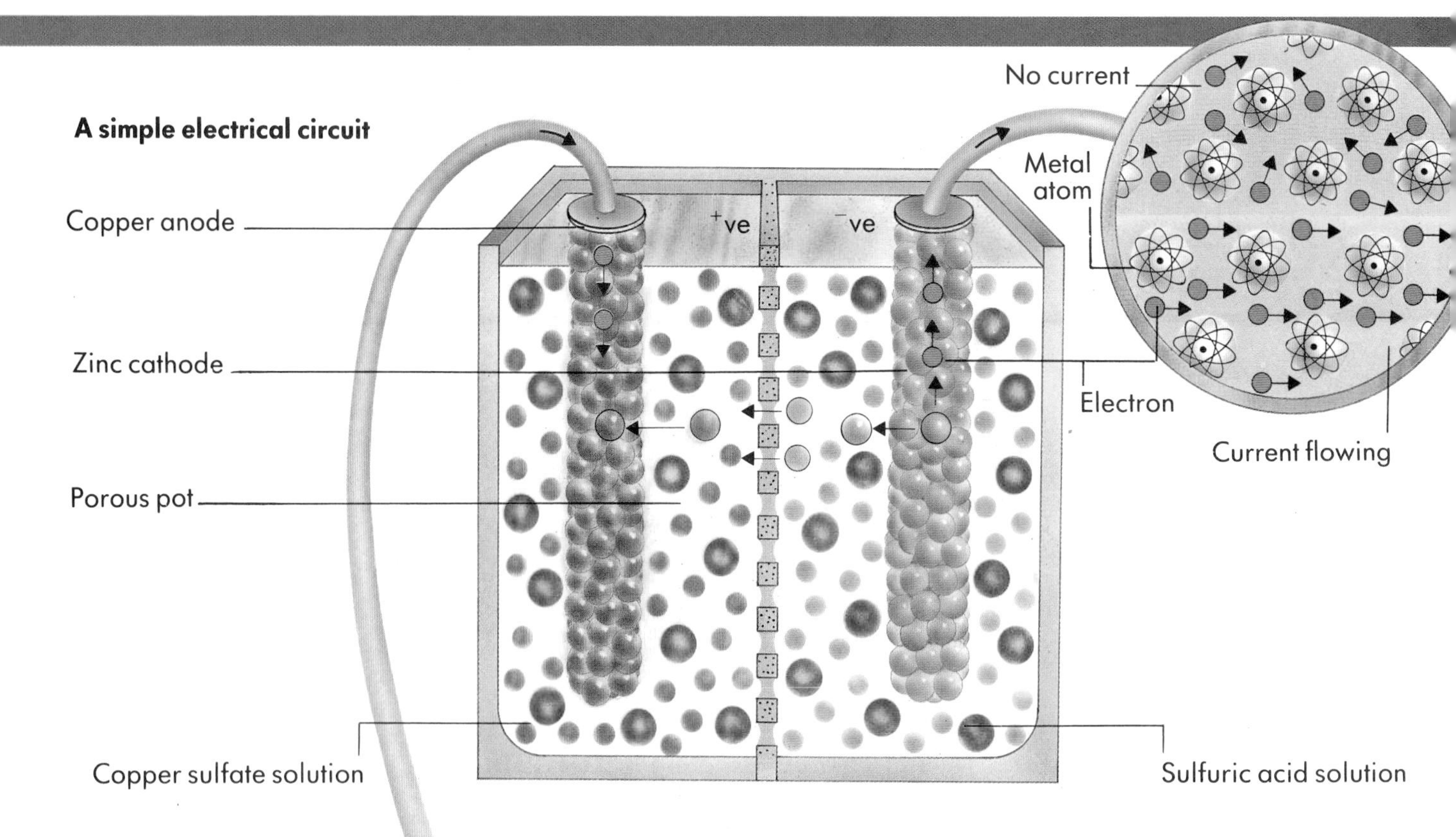

▲ A simple type of cell, or battery, called the Daniell cell. It contains a copper anode (positively charged electrode) in a copper sulfate solution and a zinc cathode (negatively charged electrode) in sulfuric acid. Atoms in the zinc cathode give up electrons, which flow through the circuit and form the current. When the electrons reach the anode, they combine with copper ions from the solution to form atoms of copper. The result is a flow of electrons from cathode to anode.

When a wire is connected to a battery in a continuous path or circuit, the electrons in the wire move along it. An electric current flows through the wire like water flowing through a pipe. But to make water flow through a pipe, a pump is needed to produce a pressure difference between its ends.

In an electrical circuit, the battery acts like an electron pump and produces an electrical pressure difference. The electrical pressure supplied by the battery is called the potential difference. It is measured in units called volts, which are named after the Italian scientist Alessandro Volta, who invented the electric battery in 1800.

The greater the voltage, the more electrons flow in the wire. We could try to measure the amount of current by counting the number of electrons that pass by. But this would be impossible because there are huge numbers of electrons in most electric currents. Instead, scientists measure electric current in amperes. One ampere is equal to a flow of about 6 quintillion electrons every second. The ampere is named for a French professor of mathematics, André Marie Ampère, who did important work on the magnetic effects of electric currents in the early 1820s.

▼ An ordinary but dangerous light bulb (here shown broken but still working) glows as the electric current flows through the filament. The thin filament has a high resistance and is made from tungsten.

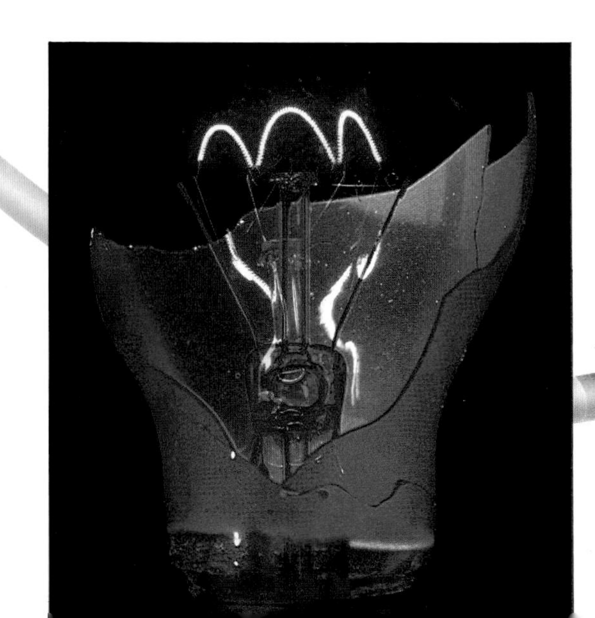

◀ In a metal wire, electrons are free to move about in any direction. As soon as a voltage is applied, as in a circuit containing a battery, the electrons move toward the positive terminal, or anode, of the battery.

▲ A space satellite, such as the *Hipparcos* star-mapping satellite shown here, uses solar cells to generate electricity from sunlight.

Ammonium chloride paste

Carbon rod

Zinc container

Manganese dioxide

Typical dry-cell battery

This "zinc-carbon" battery has a rod of carbon down its center, acting as the anode, or positive terminal. The cathode, or negative terminal, is the zinc casing. Paper soaked with ammonium chloride solution lines the casing. A chemical reaction causes the zinc atoms to produce electrons, which are able to flow along a wire connected between the terminals. A black powder of manganese dioxide and powdered carbon surrounds the carbon anode.

▼ Electrical resistance occurs when the flow of electrons is slowed down by collisions with the metal atoms or with impurity atoms in the metal. The electrons lose energy to the atoms. A circuit component with a known resistance is called a resistor.

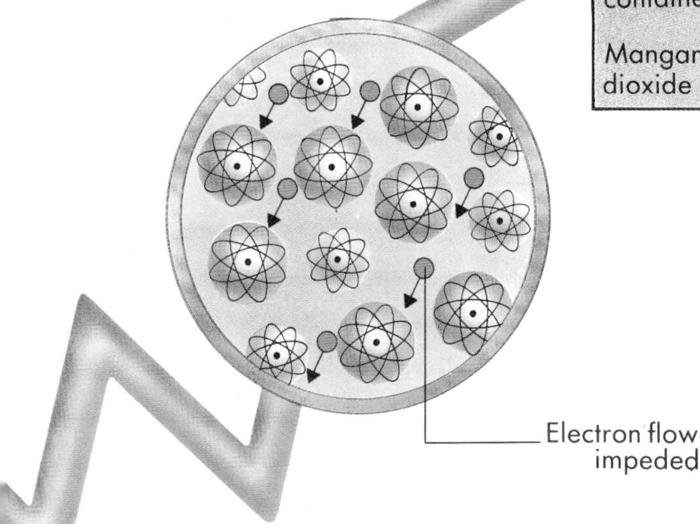

Electrons slow down as they travel through a wire, just as water slows down when it flows in a pipe. This slowing effect is called resistance. The more resistance a circuit has, the harder it is to keep the electric current flowing. A battery with a high pressure, or voltage, is needed to drive a current through a circuit with a large resistance. The amount of resistance in a circuit is measured in units called ohms, for the German scientist Georg Simon Ohm. In 1827 he discovered that the resistance of a wire is equal to the voltage divided by the current. This relationship is called Ohm's law.

Using electricity

When an electric current flows through a wire, the wire heats up. This is because of the resistance of the metal in the wire. The electrons jostle against the atoms of the metal, causing them to move. This raises the temperature of the wire because higher temperatures are linked to faster movements of atoms. The greater the resistance of the metal in the wire, the more energy the electrons lose to the atoms, and the greater the heating effect of the electric current.

In an ordinary electric light bulb, the filament is made of thin wire, because the resistance of a thin wire is greater than that of a thick wire. The filament is also made in the form of a coil. This allows a greater length of wire to be used. A longer wire has greater resistance, and therefore gets hotter and brighter.

The heating effect of an electric current is used in a fuse. A fuse is a material of fixed low-resistance so that it stops conducting if an excessive current is passed through it. If a fault develops in an electrical circuit, too much electricity may flow along the wires and heat them up. Without a fuse this could start a fire.

▼ Many children's toys use simple electric motors powered by batteries. These convert electricity into mechanical energy for movement.

Bright lights

But if there is a fuse in the circuit, the fuse quickly melts. This breaks the circuit and stops the electricity from flowing. The wiring cools down before a fire can start.

Electricity is especially useful and convenient because it can be made to do so many things. It can easily be converted into other forms of energy. A loudspeaker, for example, converts electricity into sound. Electricity can keep us cool when the weather is hot, and warm us

Colored city lights are made from long, thin tubes that glow when electricity passes through them. They are called discharge tubes. A discharge tube is filled with a vapor or gas, such as neon, at very low pressure. When the tube is switched on, electrons are emitted by electrodes at the ends of the tubes. The electrons travel along the tube, striking the atoms of the gas and causing them to emit light. The color of the light depends on the gas in the tube. Neon tubes glow bright red; argon tubes glow blue.

Fluorescent tubes are a type of discharge tube. The tube contains mercury vapor, which produces invisible ultraviolet rays when electricity flows through it. The rays fall on a fluorescent coating on the inside of the tube. This material absorbs the ultraviolet rays and reemits them as white light.

Discharge tube

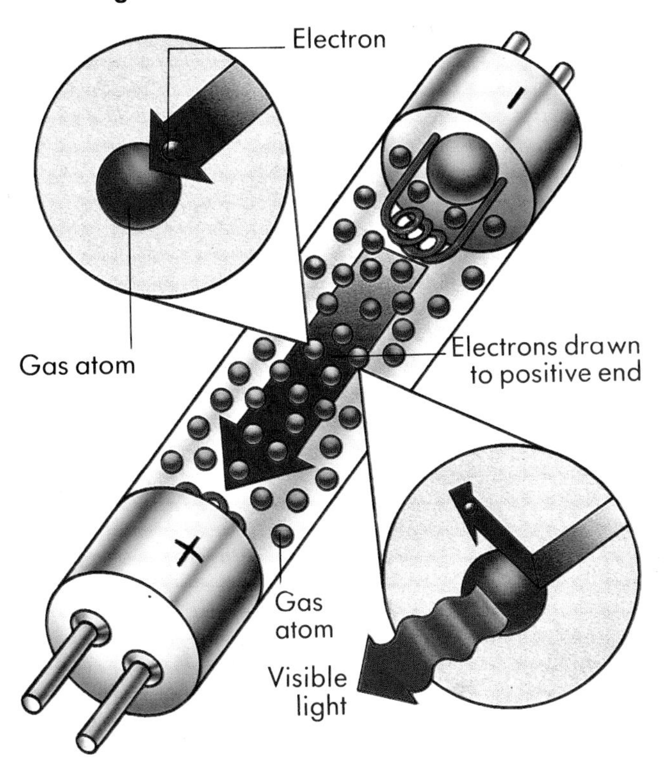

▼ Electroplating is used to coat a metal object with a thin layer of another metal. The object is hung in a tank holding a solution of a salt of the metal with which it is to be coated. A sheet of a metal is also hung in the tank. The object is connected to the cathode (negative terminal) of a battery, and the metal sheet is connected to the anode (positive terminal). The metal at the anode slowly dissolves, and the object becomes coated with the metal. All sorts of objects are electroplated, from silver coffeepots to chrome car bumpers.

Electroplating

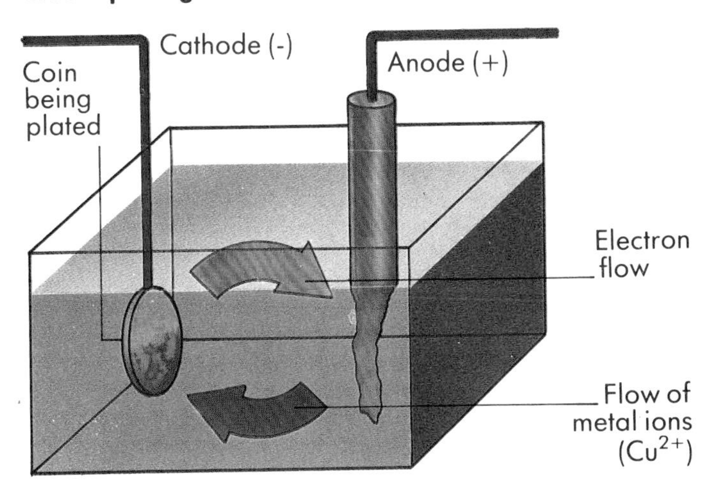

when it is cold. Electricity is vital to everyday life. Hospitals, schools, offices, and stores would come to a standstill if it were not for electricity.

Electricity can be used to break a substance down into other substances. This process is called electrolysis. Electrolysis is used in industry to extract metal from mineral ores. For example, aluminum is prepared by passing an electric current through molten aluminum oxide, prepared from the mineral bauxite.

The current is passed via electrodes through a liquid called an electrolyte. If a current is passed through a solution of common salt (sodium chloride), chlorine gas is produced at the positive electrode, or anode. Sodium is released at the negative cathode. But it immediately reacts with the water in the solution to form hydrogen gas and sodium hydroxide. Chlorine, hydrogen, and sodium hydroxide are all valuable substances made in this way.

Semiconductors

Semiconductors are materials that conduct electricity only with difficulty, unless they have been treated in some way. The most important semiconductor is silicon. Silicon is a successful conductor of electricity after minute amounts of other materials are added to it. This process is called doping. Silicon doped by adding minute amounts of phosphorus is called n-type; silicon doped by adding boron is called p-type. The effect of adding phosphorus to silicon is to add extra electrons. The extra electrons carry electricity through the silicon.

Adding boron to silicon also allows electricity to flow, but in a different way. Boron atoms have one fewer electron than silicon atoms. So when boron is added to silicon, there are places in the silicon where electrons are missing. These places are called holes. They act like positive electric charges, and carry electricity through the silicon in the same way as electrons. However, because they have a positive charge, they move in the opposite direction to the negative electrons.

Electronic engineers control the electrical properties of a semiconductor by adding precise amounts of impurities. This enables them to produce "integrated circuits." These have all the parts of an electronic circuit on a tiny silicon chip. Without them miniature televisions and personal stereos could not exist.

Inside a semiconductor

(1) Free electrons are the most important carriers of electricity in an n-type semiconductor. There may be a few holes. In a p-type semiconductor, holes are the main carriers of electricity, with a few electrons present. (2) In an n-type semiconductor, electrons are attracted to the positive terminal of the battery. In p-type material, the holes are attracted to the negative terminal. (3) When a slice of n-type material is joined to a slice of p-type material, only a small current flows when the negative terminal of the battery is connected to the p-type material. (4) When the battery is reversed, a large current flows because both holes and electrons move freely. This setup acts as a semiconductor diode, which allows current to flow in only one direction.

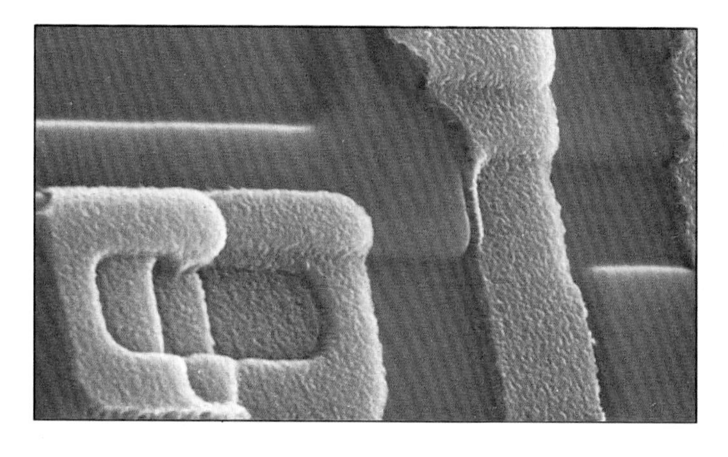

▲ A magnified view of a computer memory chip made of millions of "cells." Each cell can trap an electric charge representing a part of a number and hold it for reading.

◀ Removing silicon wafers from the doping oven. This is one of many stages in the production of silicon chips. During the doping operation a mash is applied over the wafer so that the dopants, such as boron and phosphorus, reach only certain areas.

▶ The chips on a wafer are tested and any faulty ones marked. The wafer is then cut into individual chips, and the faulty ones thrown away. The chips are put into a small plastic box, or case, with "legs," which act as connections to an external circuit. The circuits on the chip are connected to the legs by gold or aluminum wires.

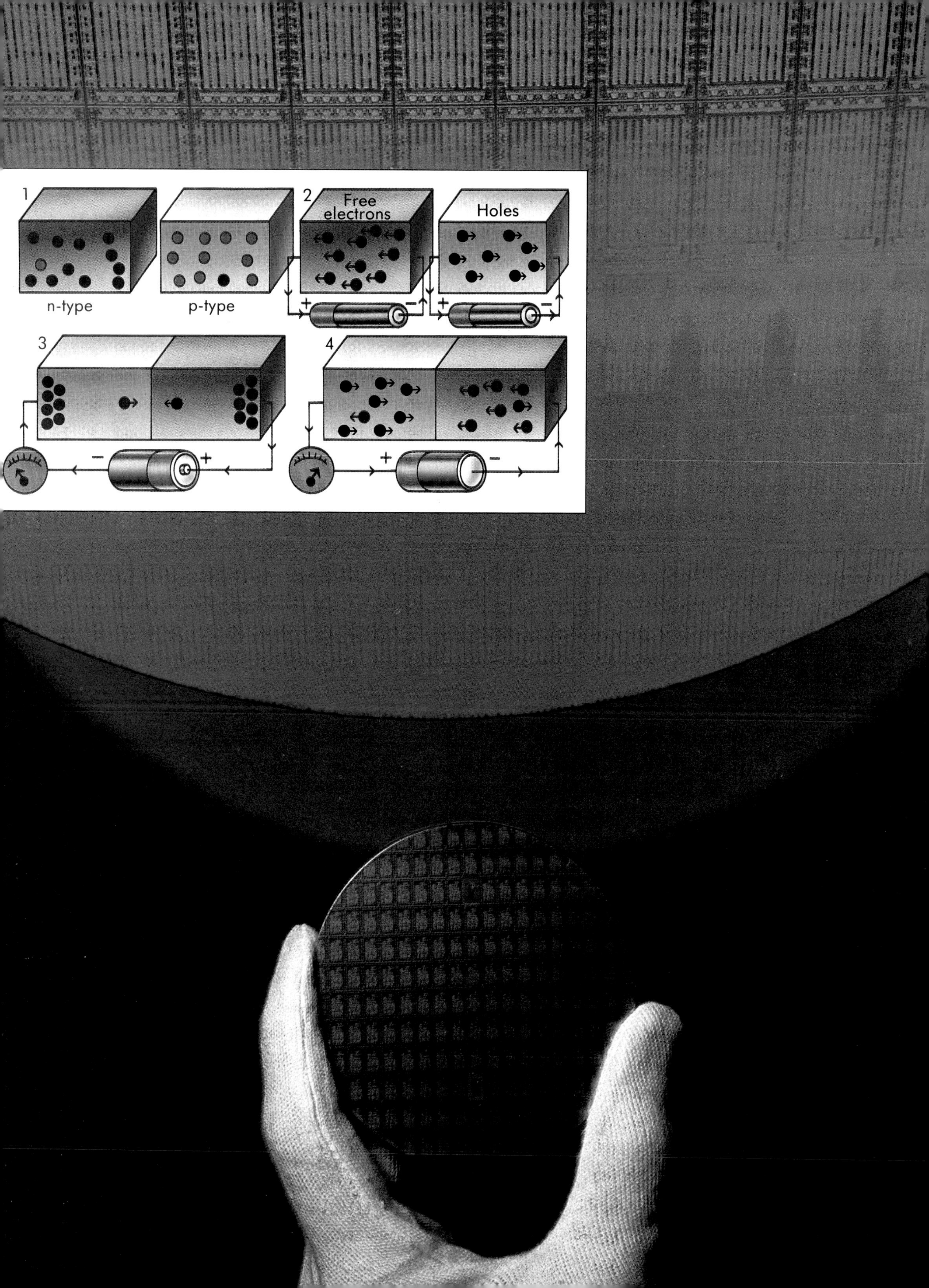
1
n-type
p-type
2
Free electrons
Holes
+
–
+
–
3
–
+
4
+
–

Magnetism

Spot facts

- *A large electromagnet at the National Magnet Laboratory, USA, gets so hot that it needs 9,000 liters (over 9,000 qt.) of cooling water every minute.*
- *The magnetic force around the giant planet Jupiter is 250,000 times greater than that of the Earth.*
- *The Earth's magnetic effects can be felt some 80,000 km (50,000 mi.) out into space.*
- *The Earth's magnetism reverses its direction from time to time. The last reversal took place about 30,000 years ago.*
- *Migrating animals probably use the Earth's magnetism to help them find the way.*

The ancient Greeks, more than 2,500 years ago, knew about magnets. They had discovered rocks which would attract small iron nails. Greek travelers told stories of mountains that could draw nails out of ships. These tales were untrue, of course, but they do demonstrate that at that time the Greeks knew about magnetism. It was not until the 16th century that the scientific study of magnets began. An English doctor, William Gilbert, pioneered the study of magnets and concluded that the Earth itself was a huge magnet. In 1820 the next advance was made by a Danish scientist, Hans Christian Oersted. He discovered that an electric current could produce magnetic effects. This form of magnetism, called electromagnetism, is used in electric motors and devices such as doorbells and telephones.

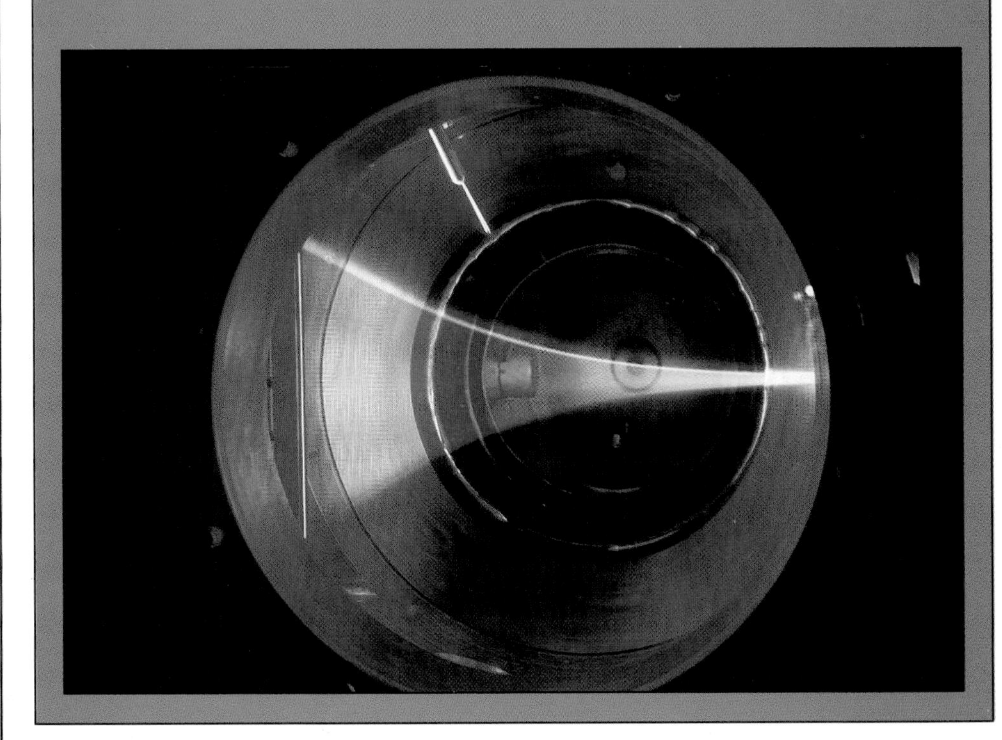

► A huge electromagnet shapes a blue haze of hot hydrogen inside a device which attempts to harness the power of nuclear fusion. This is the process which produces the Sun's energy. A strong magnetic field is used to contain the very hot material that is produced by the fusion process.

Magnets

Magnets attract things made of iron or steel. Some other metals, such as cobalt and nickel, are also attracted to a magnet. But most metals, such as gold, copper, and aluminum, are not attracted in the same way. Neither are plastics, paper, cloth, or glass attracted to magnets. These things are said to be nonmagnetic.

Iron filings stick to a magnet most strongly at two points, usually near its ends. It is at these points, called the poles, that the magnet's power is strongest. One of them is called the north pole; the other is called the south pole. If a bar magnet is hung at its center on a thread so that it can swing freely, the north pole points north. This effect is used in a compass. All planes and large ships are fitted with a compass as a vital part of the navigation system.

If you put the north pole of one magnet near the north pole of another, the magnets push apart, or repel, each other. Two south poles behave in the same way, also repelling each other. Magnets attract each other only if different poles are close together. Scientists say: "Like poles repel, unlike poles attract."

A giant magnet

The Earth acts as if it were a giant bar magnet, and has two magnetic poles. The Earth's north magnetic pole is in the Canadian Arctic, about 1,600 km (1,000 mi.) from the true, or geographic, North Pole. The south magnetic pole is in Adélie Land about 2,400 km (1,500 mi.) from the geographic South Pole in central Antarctica. The magnetism of these poles makes the north pole of a compass needle point north, and the south pole of a compass point south.

But because the Earth's magnetic and geographical poles are in slightly different locations, a compass does not point exactly due north or south. The slight difference between the magnetic and geographical poles is called the magnetic variation. It changes all the time. Navigators and map readers must allow for it when finding their way.

The Earth's magnetism is produced by molten metal deep within the Earth's core. As the Earth spins, electric currents are created in the molten metal. These currents produce the Earth's magnetic force.

Earth's magnetism

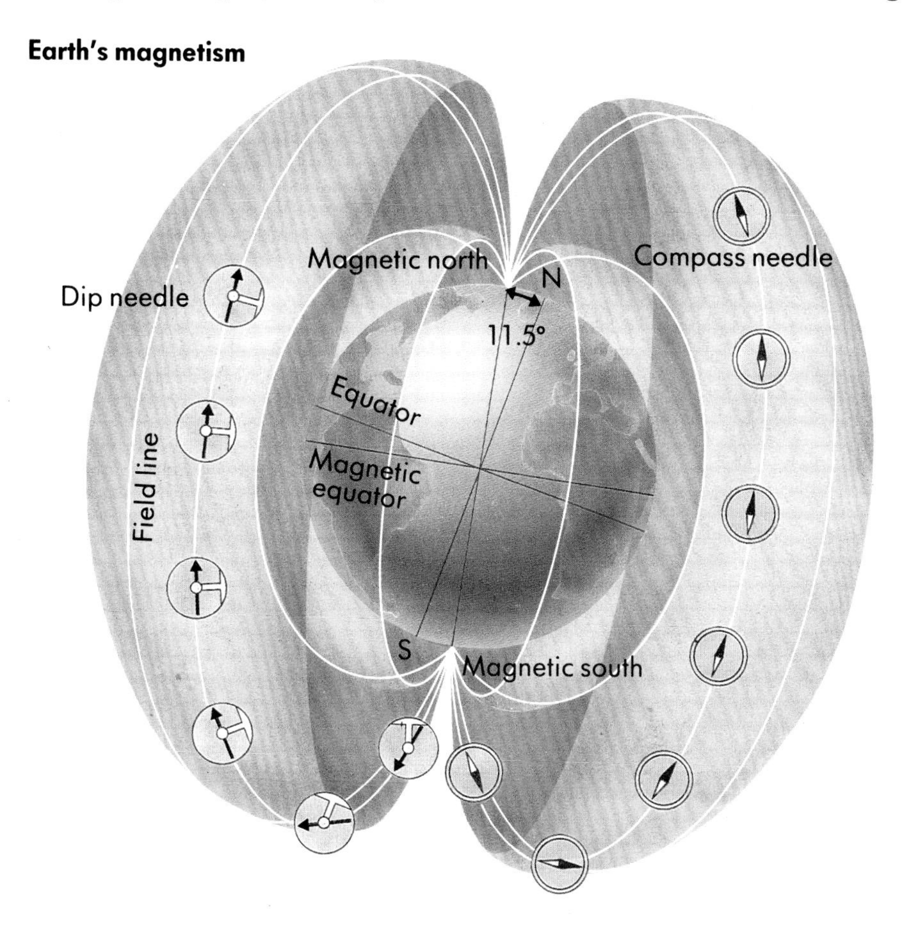

▲ Iron filings cluster around the poles of a magnet, where the magnetic force is strongest.

◀ We live on a huge magnet! Magnetized needles that are free to rotate point more or less toward the Earth's magnetic poles.

Magnetic field

The space around a magnet, where the magnetic force can be detected, is called a magnetic field. A kind of map of a magnetic field can be made with iron filings. Put a magnet under a piece of white paper and sprinkle iron filings on top. Tap the paper gently. You will see the filings form a pattern of lines. These lines follow the lines of force of the magnet.

The lines of force are close together near the poles, where the effect of the magnet is concentrated. Away from the poles, the magnetic effect is weaker and the lines of force are farther apart. The lines show the force a small north pole would experience in the field.

How are magnets made? One way is by stroking a piece of steel, such as a thin needle, with a magnet. To understand why this happens remember that the needle, like all other things, is made up of atoms. In a magnetic material, such as iron or steel, the atoms are like miniature magnets.

The atomic magnets are grouped together into small areas called domains, which are like mini-magnets inside the material. When an iron needle is stroked with a magnet, the domains become lined up so that they all point in the same direction. In this way the needle has become magnetized.

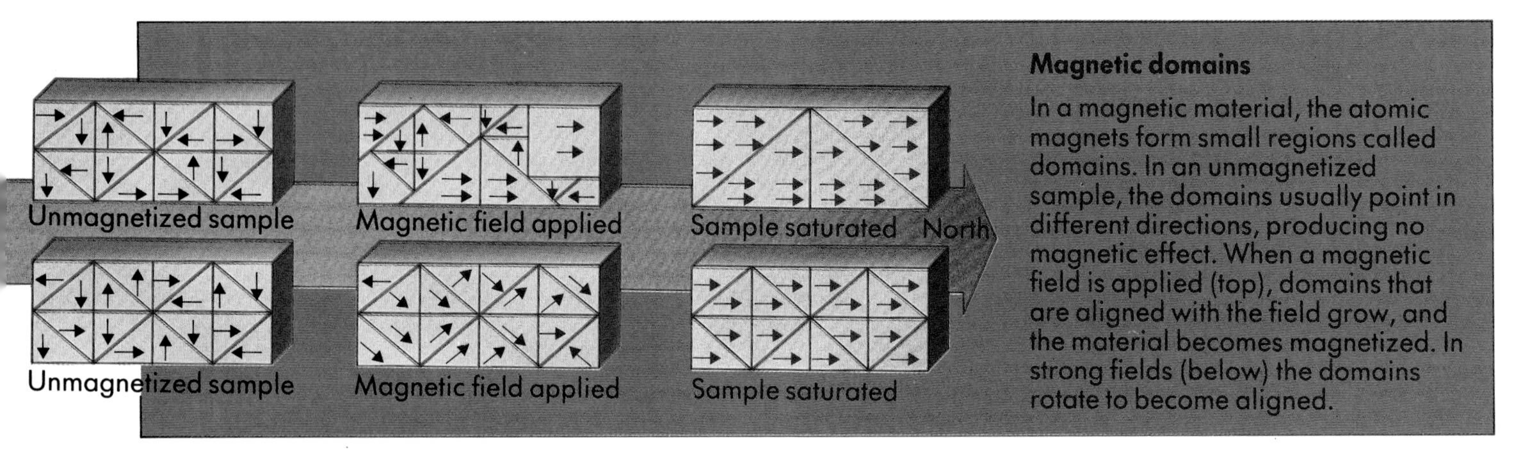

Magnetic domains

In a magnetic material, the atomic magnets form small regions called domains. In an unmagnetized sample, the domains usually point in different directions, producing no magnetic effect. When a magnetic field is applied (top), domains that are aligned with the field grow, and the material becomes magnetized. In strong fields (below) the domains rotate to become aligned.

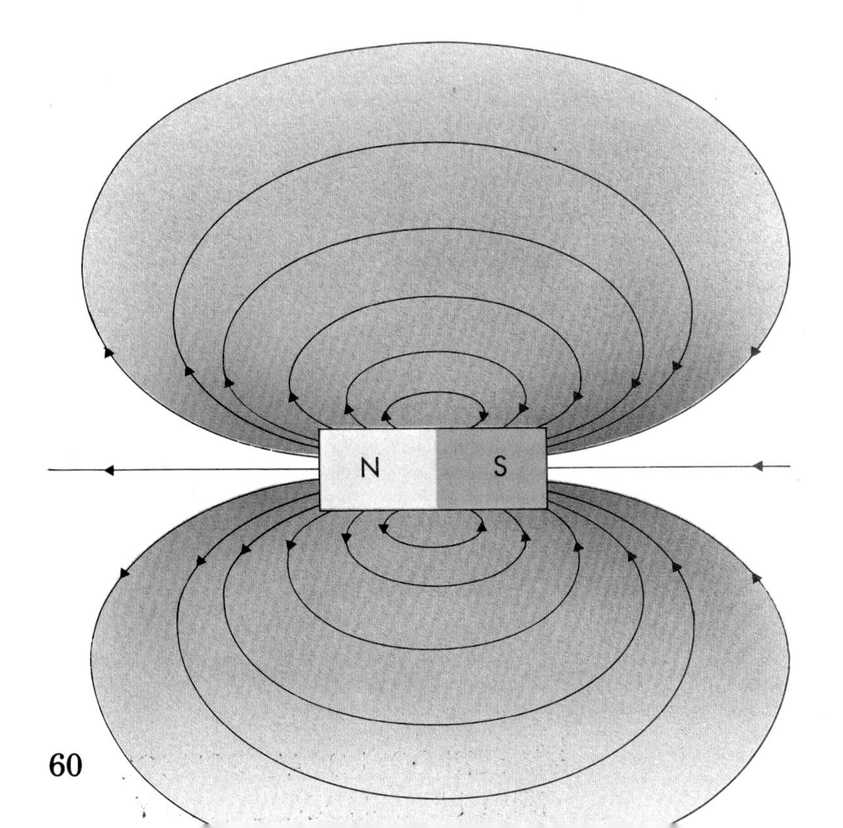

◀ ▼ A magnetic field can be represented by lines, called lines of force, linking north and south poles. The arrows give the direction of the force on an imaginary north pole. Similar magnetic poles repel each other, giving a "neutral point" between them where there is no force. Unlike poles, on the other hand, are attracted to each other, with lines of force filling the space between them.

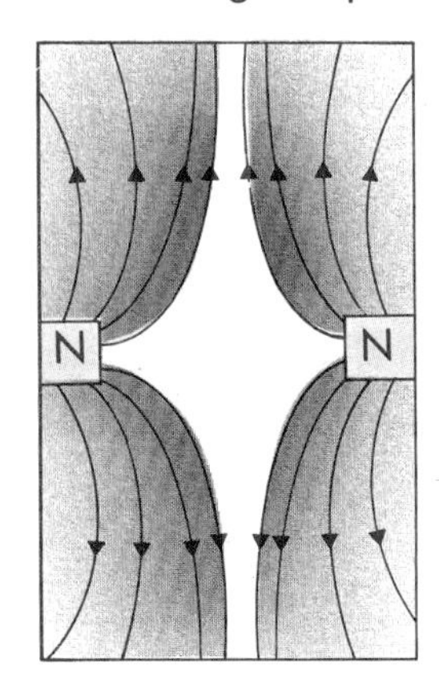

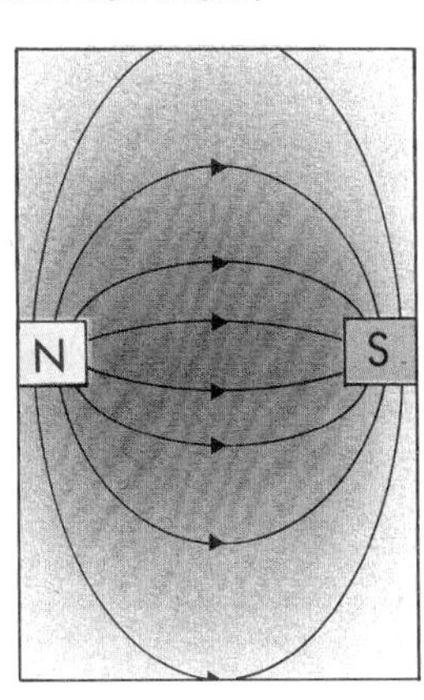

Electricity and magnetism

Moving electric charges – an electric current – cause magnetism. Indeed, magnetic fields are not truly different from electric fields. Rather, both are aspects of the same thing, the "electromagnetic field."

In 1820 the Danish physicist Hans Christian Oersted noticed that a current flowing in a wire deflected a compass needle nearby. It did not take long to discover that the magnetic field around a wire carrying a current is complicated. The strength of the field depends on several things — the strength of the current, the length of the wire, and the distance to the wire.

In some instances, the field has a relatively simple form. Around a straight wire, the lines of force are circles. Around a solenoid – a coil with many turns – the field resembles that of a bar magnet. It has a north pole at one end and a south pole at the other.

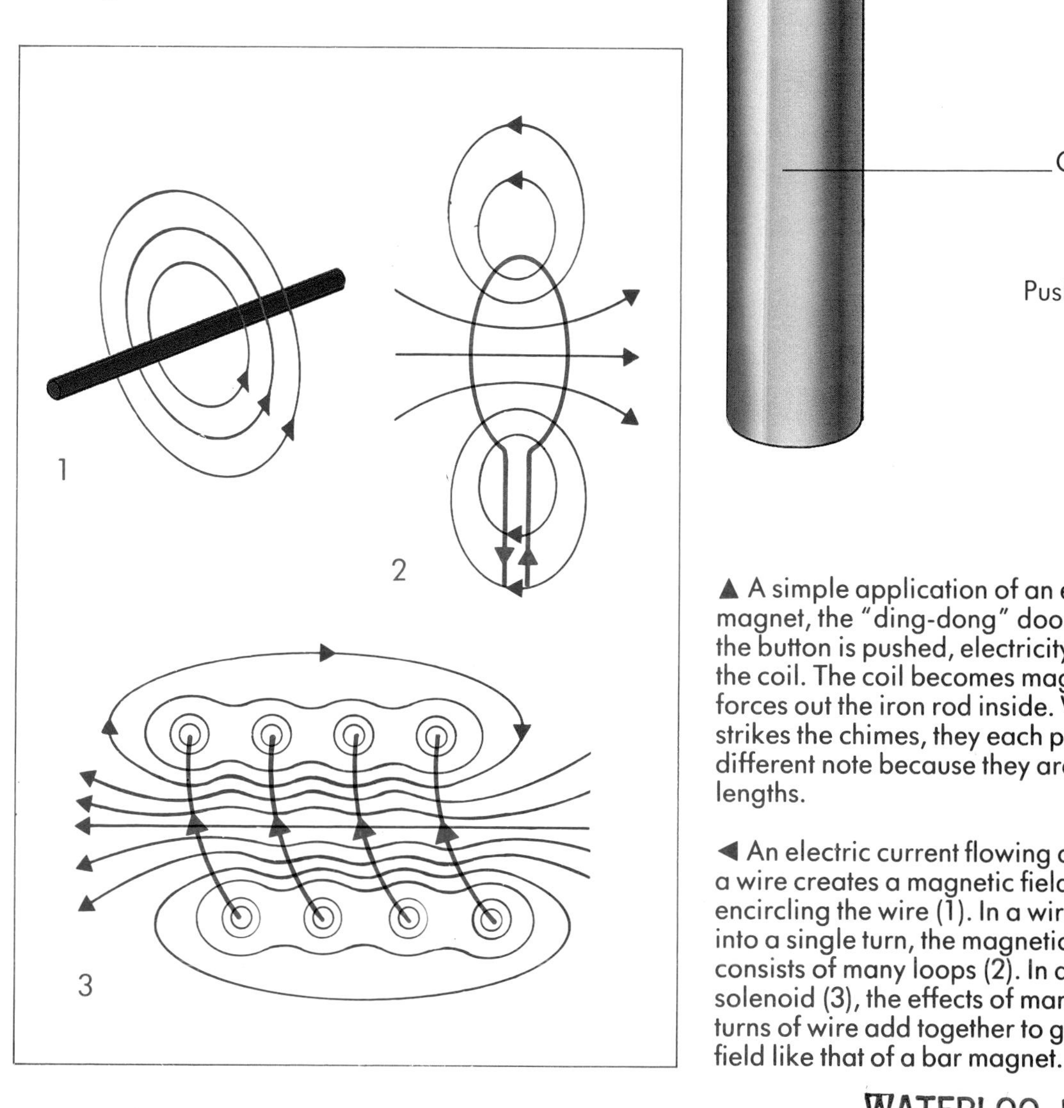

Electromagnetic doorbell

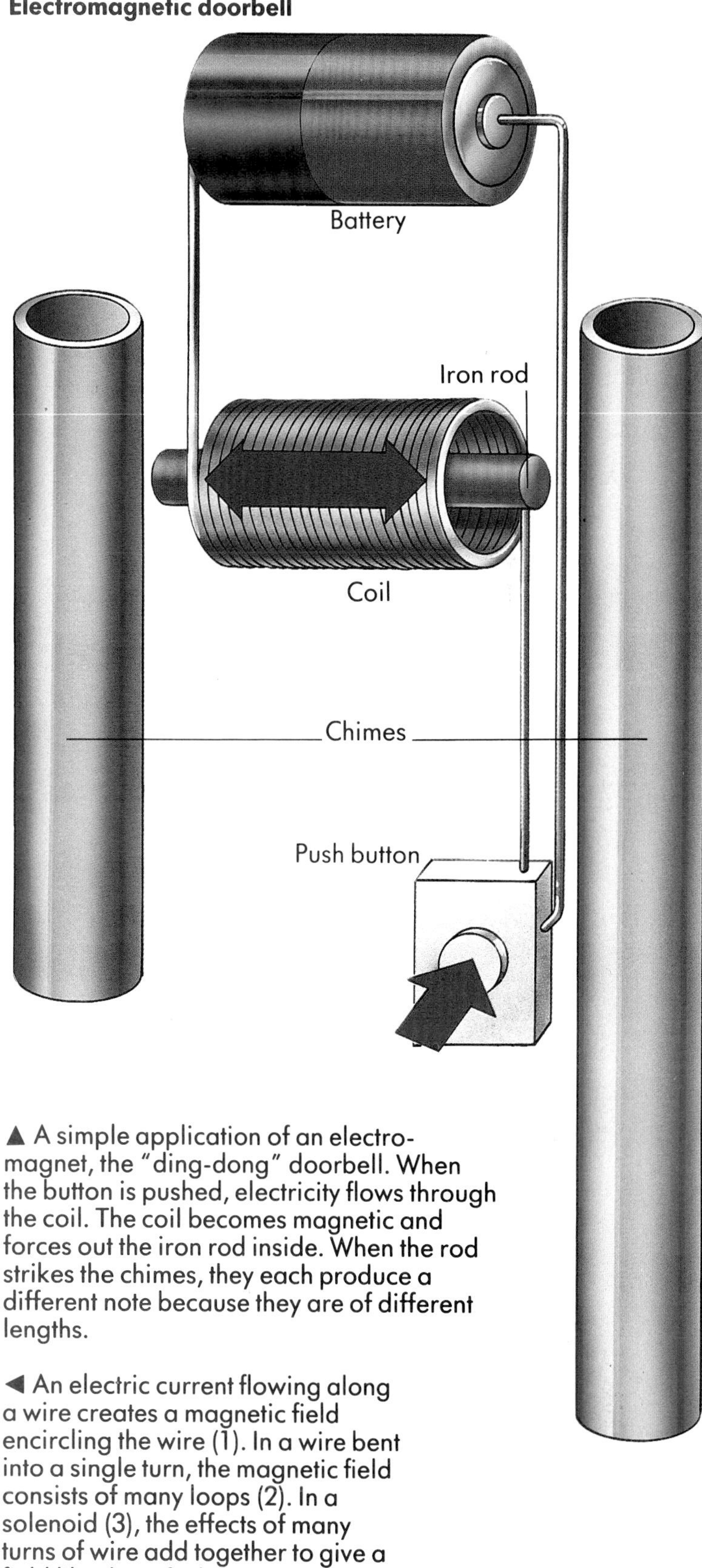

▲ A simple application of an electromagnet, the "ding-dong" doorbell. When the button is pushed, electricity flows through the coil. The coil becomes magnetic and forces out the iron rod inside. When the rod strikes the chimes, they each produce a different note because they are of different lengths.

◀ An electric current flowing along a wire creates a magnetic field encircling the wire (1). In a wire bent into a single turn, the magnetic field consists of many loops (2). In a solenoid (3), the effects of many turns of wire add together to give a field like that of a bar magnet.

Electromagnetism

The first electromagnet was made in 1823 by an Englishman, William Sturgeon. He wound an insulated copper wire into a coil around an iron bar. When a current flowed through the coil, the bar became a strong magnet. The strength of an electromagnet depends on the number of turns of wire in the coil and on the strength of the current flowing through it. The more turns and the larger the current, the stronger the electromagnet becomes.

Superconducting electromagnets

Today, the strongest electromagnets are made using superconducting coils. The coils are made of materials that lose all electrical resistance when they are cooled to very low temperatures, around −269°C (−450°F). They can carry the large electric currents needed in powerful electromagnets. Superconducting magnets are used in particle accelerators, which are used for atom-smashing experiments.

Superconducting magnets are also used in medicine. They are an essential part of a type of body scanner, called an MR scanner. These scanners use a process called magnetic resonance (MR) imaging to produce detailed pictures of the inside of a patient's body.

The patient lies in a strong magnetic field produced by an electric current flowing in a superconducting coil. Radio signals are beamed into the area of the body being investigated. The nuclei of the atoms of the body produce tiny magnetic signals, which are picked up by detectors. A computer is used to form a picture of the inside of the body from these magnetic signals.

Electromagnets are found in many household appliances. In a television set, electromagnets are used to control the beams of electrons that form the pictures on the screen. In a telephone, an electromagnet moves the plate in the earpiece that produces the sounds we hear. In a loudspeaker, electric currents produced by a record player or tape recorder are converted into sounds by an electromagnet. Electromagnets in the form of motors are also found in vacuum cleaners, food mixers, hair dryers, and washing machines.

► A magnetic levitation train is held in the air by superconducting electromagnets. These magnets produce strong magnetic fields because large currents can flow through them without resistance.

▼ Very powerful electromagnets are often used in junkyards. The magnetism does not exist when the current is switched off, and the magnet releases its load.

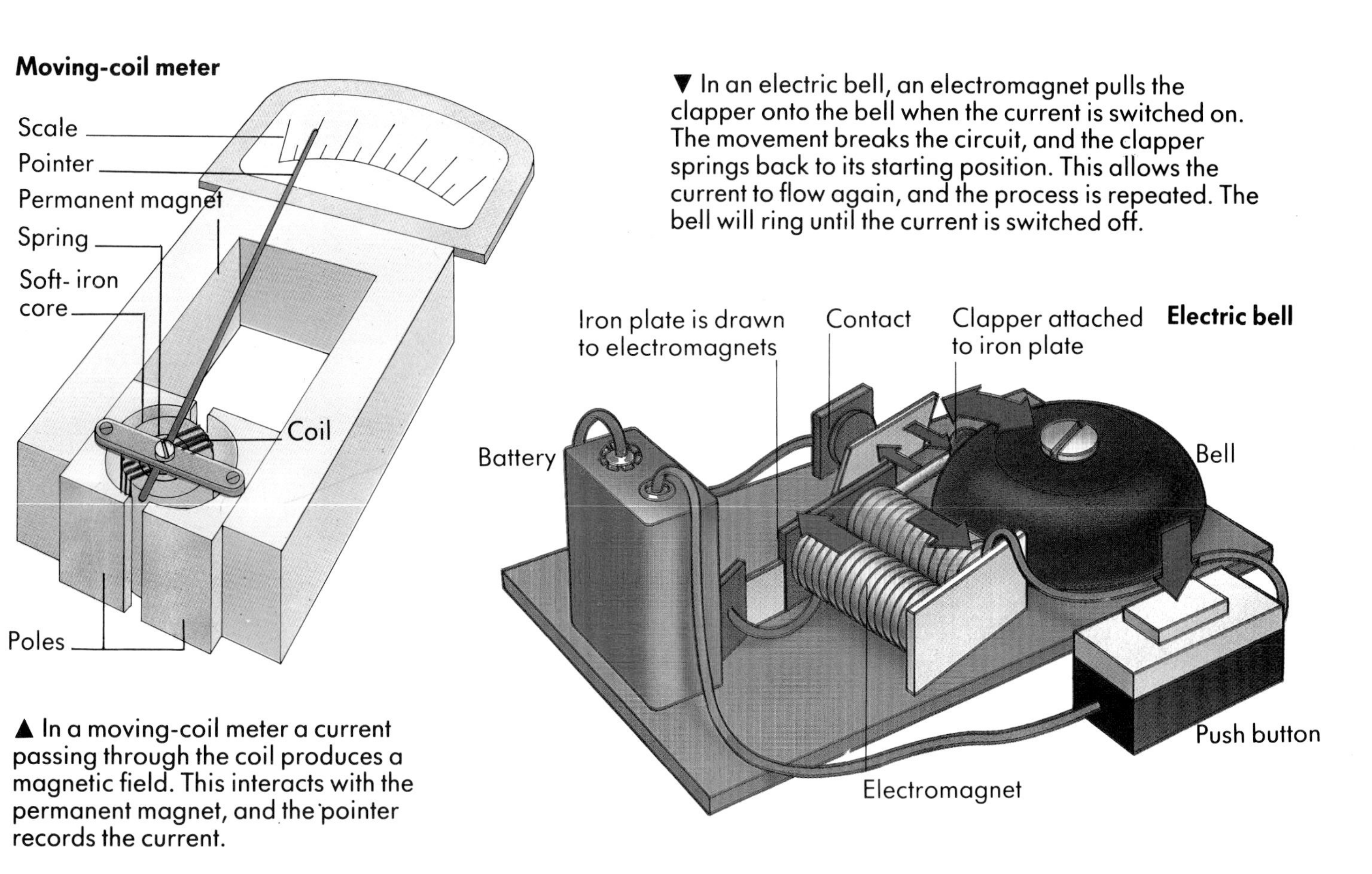

▼ In an electric bell, an electromagnet pulls the clapper onto the bell when the current is switched on. The movement breaks the circuit, and the clapper springs back to its starting position. This allows the current to flow again, and the process is repeated. The bell will ring until the current is switched off.

▲ In a moving-coil meter a current passing through the coil produces a magnetic field. This interacts with the permanent magnet, and the pointer records the current.

▼ Circuit breakers are switches used in power stations. The magnetic effect of an excess current opens the switch, preventing the circuit from becoming overloaded.

► A transformer consists essentially of two electromagnets. By altering the number of turns in the coils, it can be used to increase or decrease the input voltage.

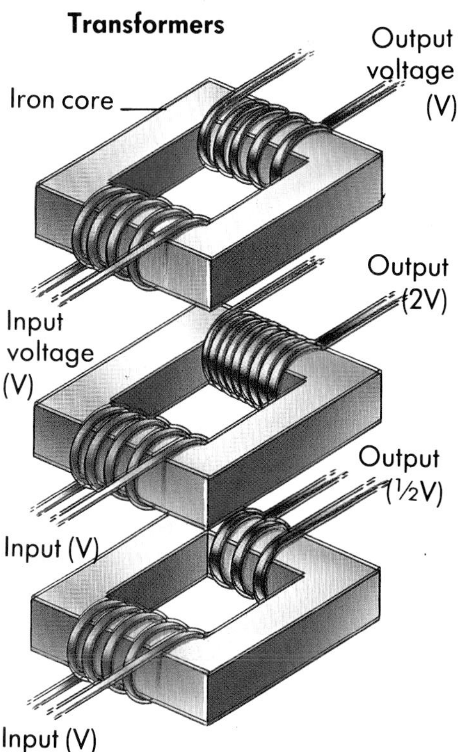

Generators and motors

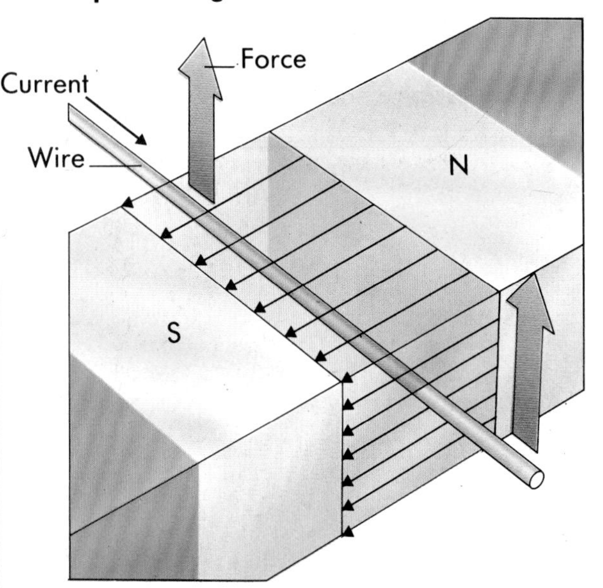

When a wire moves in a magnetic field, a current flows if the wire is part of a circuit, because the electrons in the wire experience a force. This is the principle that underlies the operation of a generator. When a current flows in a wire in a magnetic field, the wire experiences a force. This is how a motor works.

Electrical generators and motors are related in the way they work. A generator converts energy of movement into electrical energy. An electric motor does the opposite; it converts electrical energy into energy of movement. The common factor between the two pieces of equipment is the effect of a magnetic field on moving electrons.

The electrons in a wire moving through a magnetic field experience a force, which sets the electrons moving along the wire. This is how a generator works. If the wire is in the form of a loop being spun in the magnetic field, the current produced moves first in one direction and then in the opposite direction. This happens because the two sides of the coil move alternately up and down through the field. This sort of current, which moves back and forth, is called alternating current (AC). It is the kind of electric current supplied by the power lines found in our homes.

It is possible to produce one-way electric current, or direct current (DC), from a rotating coil. A split ring, called a commutator, is attached to the ends of the coil. The commutator connects the coil to the circuit.

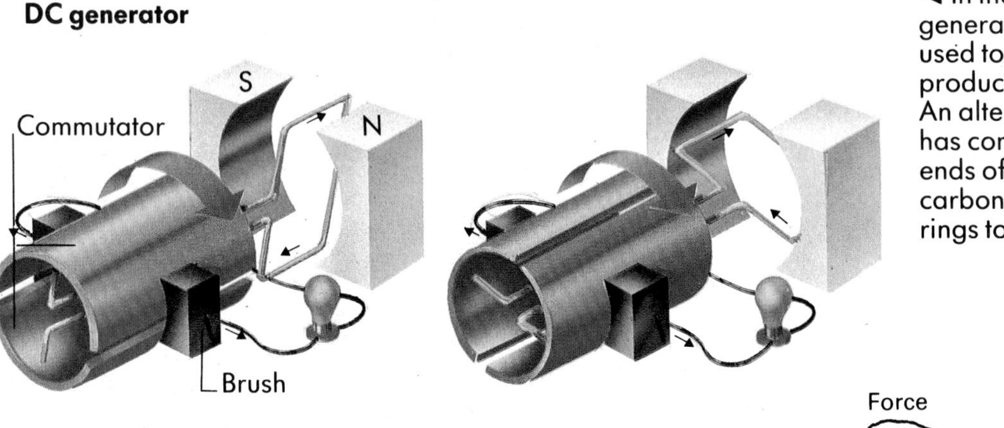

◀ In the direct-current (DC) generator, a split-ring commutator is used to ensure that the current produced flows in only one direction. An alternating-current (AC) generator has complete rings connected to the ends of the coil. In both cases, carbon brushes press against the rings to draw off the current.

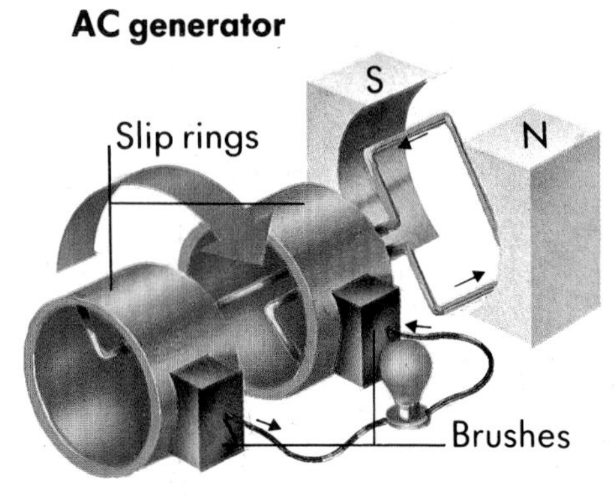

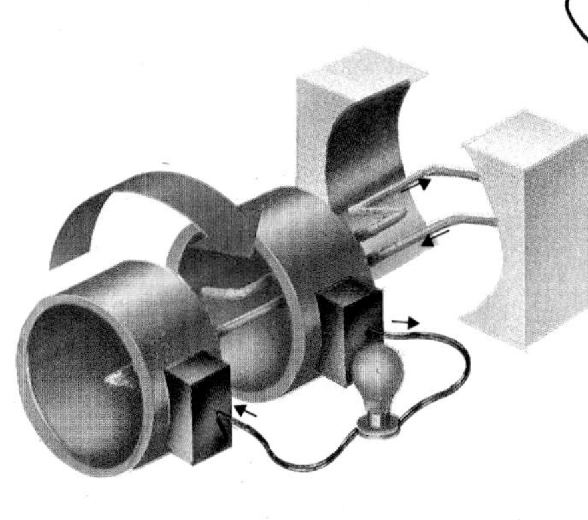

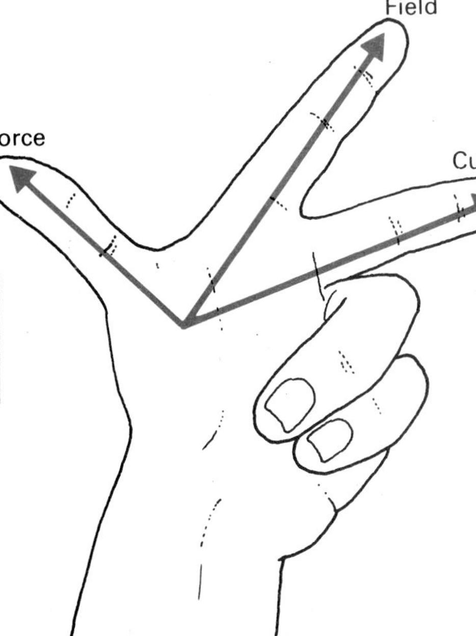

In an electric motor, a current is set up through a wire placed in a magnetic field. The moving electrons then experience a force due to the magnetic field, which makes the wire move. If the wire is in the form of a loop, the forces acting on the two sides of the loop make it spin.

A simple motor requires an alternating current to make it work. As the coil turns, the current is reversed at the right moment to continue the rotation. In a direct-current motor, a commutator reverses the current through the coil at each half-rotation, and in this way keeps the rotation going.

There are various kinds of electric motors. They are used in factories, electric railroads, and household appliances. Most household appliances use a motor in which the magnetic field is produced by an electromagnet. The electromagnet is connected to the same electrical supply as the rotating coil. To increase the power, there are many coils in these motors. Each coil is in a slightly different position from its neighbors. The commutator is split into many segments, one for each coil. The motors are called universal motors, because they can run on both direct and alternating current.

▲ A powerful electric motor is used to drive a winch at a mine in Zimbabwe. The winch winds in a steel cable to hoist rock from deep down the mine.

▼ Fleming's left- and right-hand rules. They indicate the directions of current (opposite to electron flow), magnetic field, and movement for motors, (left-hand) and generators (right-hand). The index finger points in the direction of the field, the thumb in that of the motion, and the middle finger in that of the current.

▼ In a direct-current motor, the commutator, which connects the coil to the current, reverses the direction of the electric current after the coil has turned half a turn. As a result, the coil keeps turning in the same direction. Without a commutator, the coil would come to rest after half a turn, with the coil horizontal.

▼ An alternating-current motor does not need a commutator because the current in the coil is continually reversing in direction. Instead it uses a pair of slip rings, one connected to each end of the coil. The coil rotates at a speed which keeps it in step with the changes in the direction of the current.

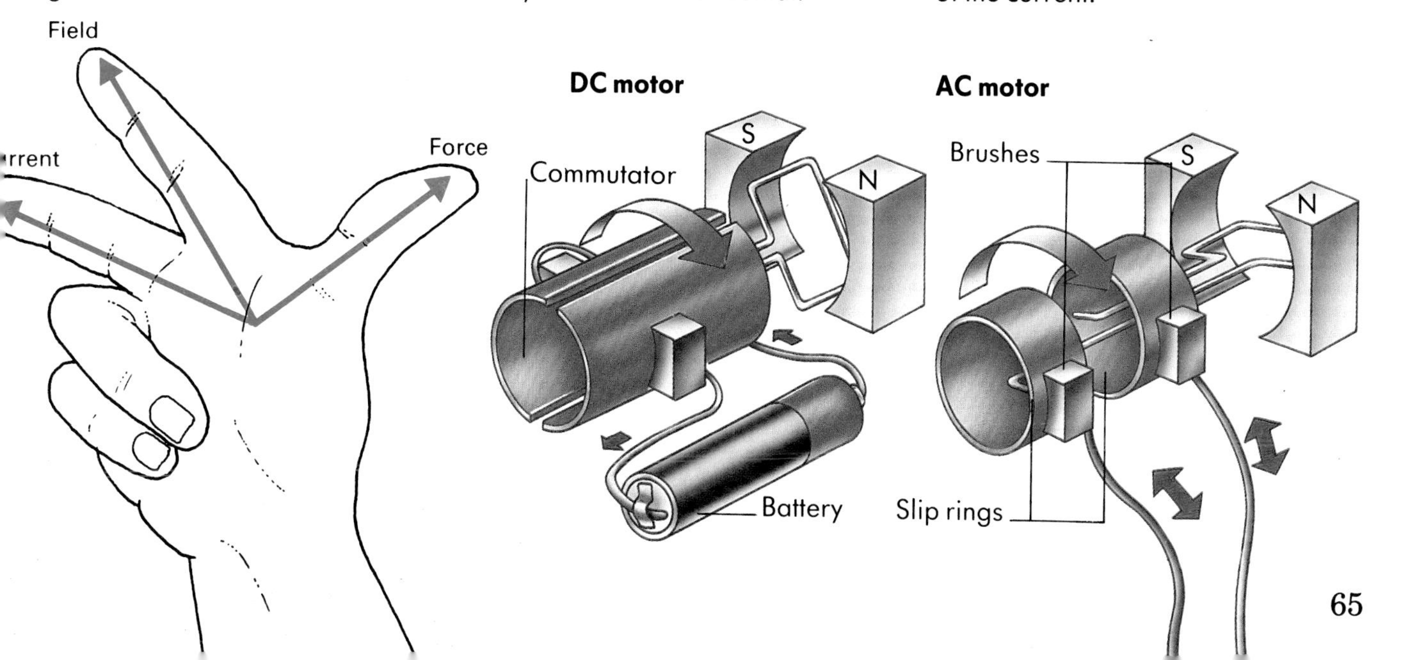

Light and radiation

Light shining through a hole forms a beam, or ray, that travels in a straight line. Light rays can be seen to bounce off mirrors, and bend when they enter a transparent material. But this is not the whole story. Light is also a form of wave, an electromagnetic wave. Furthermore, modern developments in the use of light, such as the laser, can be explained only if light also consists of packets of energy, or particles of light, called photons.

Spot facts

- *Light travels at a speed of 300,000 km/s (186,000 mps). It takes one-tenth of a second to travel from New York to London, England, 8½ minutes to reach the Earth from the Sun, and 4⅓ years to reach Earth from the next nearest star.*
- *Albert Einstein, that most famous scientist, showed in 1905 that nothing could travel faster than the speed of light.*
- *Laser light beams shined from the Earth have been reflected off mirrors left by astronauts on the Moon. These experiments have been used to measure the precise distance to the Moon.*

► A simple experiment that reveals much! When light passes through a prism, two phenomena occur. The beam bends away from its original direction. This bending is called refraction. The beam also disperses into a band of colors known as a spectrum. This shows that white light is a mixture of these colors.

Reflection and refraction

One of the earliest discoveries about light was that it seems to travel in straight lines. Other discoveries concerned what happens to a light ray when it meets a surface, for example the surface of a mirror or a glass surface.

When a beam of light falls on a shiny surface, it bounces off, or is reflected. The law of reflection states that the angle at which the ray strikes the surface, called the angle of incidence, is always equal to the angle at which the light beam leaves it. Simple diagrams using this law demonstrate how a mirror forms an image. The reflected image appears to come from behind the mirror.

Light rays bend when they enter a transparent material. This bending is called refraction. Refraction explains why it is so hard to spear a fish from the bank of a river. The fish is not where it seems to be, because the light from the fish is refracted as it leaves the water.

Refraction also explains why a drinking straw appears to be bent where it dips into a glass of water. Refraction occurs because light travels more slowly in water and glass, which are denser than air. This causes a light ray to swerve as it enters these materials, in the same way that a racing car swerves if it drives off the track onto a rougher surface.

◀ The John Hancock Tower in Boston. The glass in many modern buildings acts like a giant mirror, often creating startling effects. The images in a mirror or reflected in a glass window are reversed left-to-right.

▼ The image of a tree reflected in a lake is upside-down and the same size as the real tree appears to be.

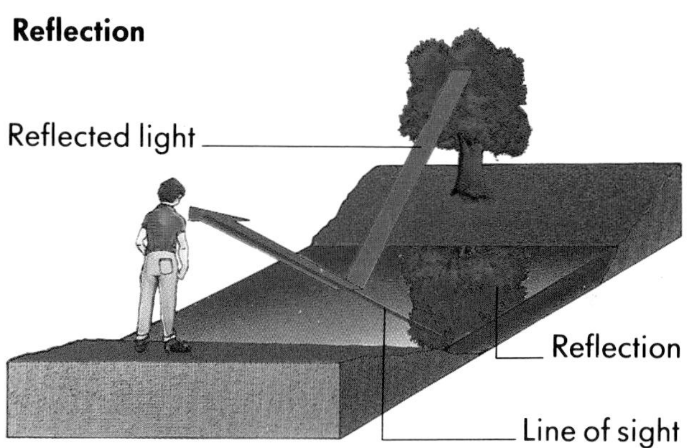

▼ A mirage of a tree in a desert is formed because light from a distant tree is refracted when it passes through the hot air near the ground.

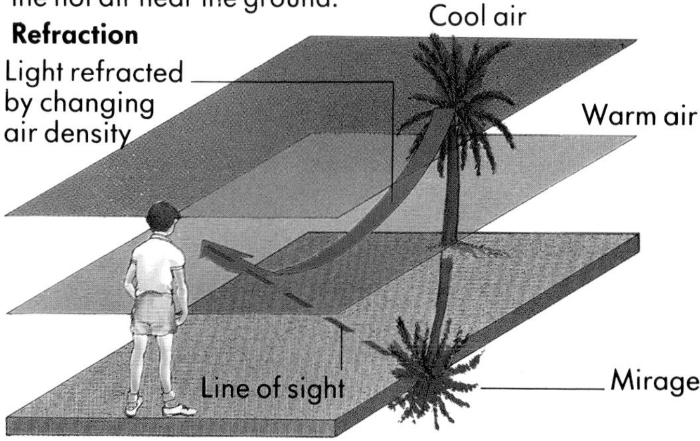

Curved mirrors and lenses

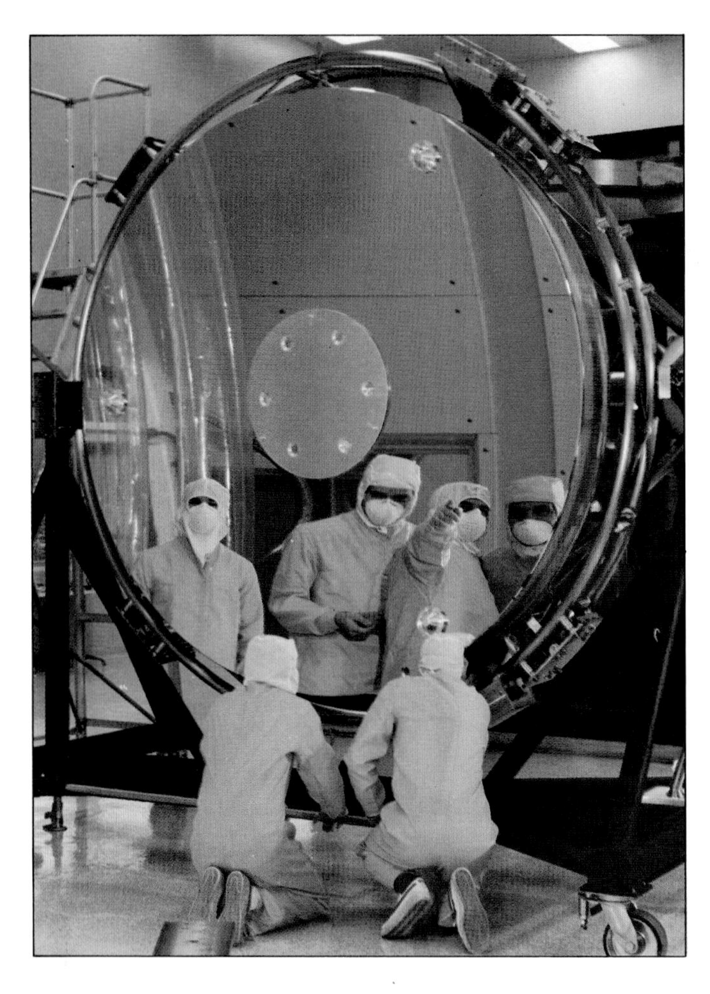

▲ Large astronomical telescopes generally use curved mirrors. This one was for the ill-fated Hubble Space Telescope, which was launched into orbit in 1990. Early telescopes used lenses, but produced poor images. Isaac Newton built the first reflecting telescope in 1671. Mirrors can be made much larger than lenses.

Mirror images

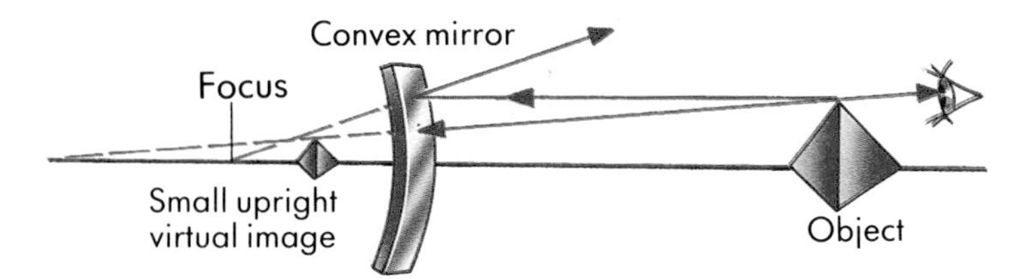

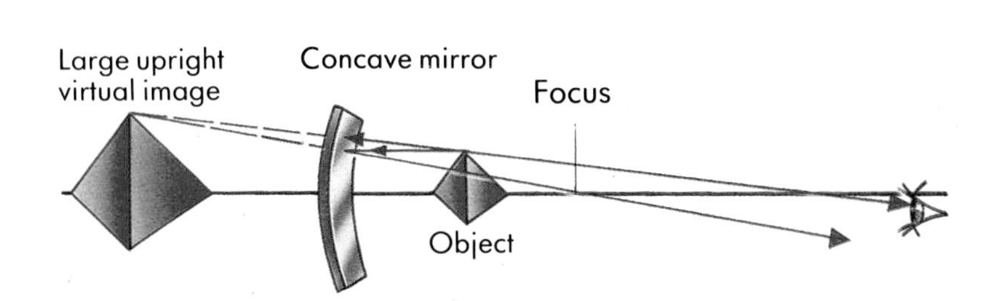

◄ A convex mirror produces a small, upright virtual image behind the mirror. A concave mirror produces a large upright, virtual image if the object is between the focus and the mirror. If the object is farther away, a small inverted (upside-down) real image is formed in front of the mirror.

There are two kinds of curved mirrors: concave mirrors that bend inward, and convex ones that bulge outward. Concave mirrors are used in astronomical telescopes, shaving mirrors, and car headlight reflectors. Convex mirrors are used for car rearview mirrors and in supermarkets and other stores to give the staff a good view of the store floor.

If a beam of parallel light rays falls on a concave mirror, the rays are reflected so that they all pass through a single point in front of the mirror. This point is called the focus. Light falling on a convex mirror is spread out as if it were coming from behind the mirror.

Curved mirrors can produce images, or pictures, of objects placed in front of them. The size of the image and its position depend upon where the object is placed. If the object is far away from a concave mirror, no image can be seen in the mirror. However, if a piece of paper is held in front of the mirror, a small, upside-down image can be seen on the paper. This is called a real image. If an object is close to a concave mirror, a large upright image of it can be seen in the mirror. This kind of image, which can be seen in the mirror but not focused on to a piece of paper, is called a virtual image.

Lenses are pieces of glass or other transparent material with curved sides. They are used in cameras, small telescope, and eyeglasses. Convex lenses have sides that bulge outward. Concave lenses have sides that bend inward. Convex lenses can bring parallel light to a focus at one point, the focus. A concave lens makes parallel light diverge away from its focal point. Like mirrors, lenses can form images. The type and size of image depend on which type of lens is used and where the object is placed.

▲ A magnifying glass is a convex lens. An object between the lens and its focus appears larger, or magnified, but can appear farther away from the lens than it really is.

◀ A lighthouse uses both mirrors and lenses to produce a powerful beam that can sweep the horizon. A solid lens would be too big and too heavy. Instead, Fresnel lenses, named after a French scientist, are used. In such a lens a flat surface is divided into a series of circles. Relatively thin ridges of glass are set in each circle at angles that would be found in a solid lens at that point.

Convex and concave lenses

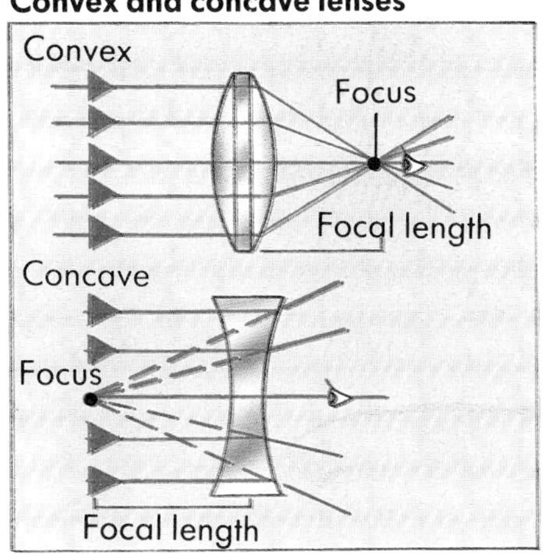

Convex lens image

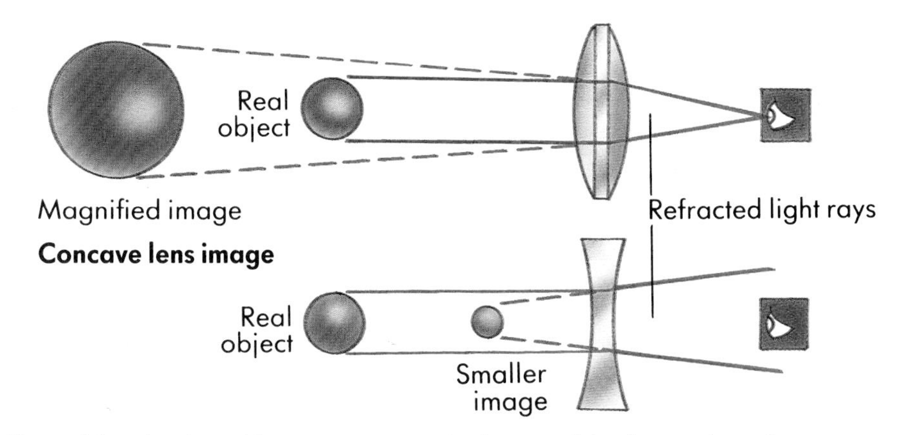

▲ A parallel beam of light entering a concave lens is bent to a single point known as the focus. On leaving a concave lens parallel light rays spread out as if from a point behind the lens.

▲ If an object is placed between a convex lens and its focus, an enlarged virtual image is seen through the lens. A concave, or diverging, lens always produces a smaller virtual image between the lens and the focus. A real image is one where the light rays actually come from the object. With a virtual image the rays only appear to come from the object and cannot be focused onto a screen.

Waves of light

What is light? Many scientists have studied light and tried to answer this question. In the late 17th century, the great English scientist Isaac Newton suggested that a light beam consisted of a stream of tiny particles, which he called corpuscles.

In about 1690 a Dutch physicist, Christiaan Huygens, put forward another idea. He thought that light traveled in waves. If you drop a pebble into water, you will see waves, or ripples, move out from the point where the pebble hits the water. Huygens believed that light traveled in a similar way, but that the light waves were very small. He thought that the distance between the tops of any two neighboring light waves was only a few ten-thousandths of a millimeter (a few hundred thousandths of an inch).

Although Huygens's theory explained the properties of light, such as reflection and refraction, it took many years for this theory to win support. In 1801 the Englishman Thomas Young performed an experiment which showed that light consisted of waves.

In Young's experiment, light from a small point source was shined onto a screen through two fine slits side by side. The light source was a sodium lamp, which gave light of a pure color. Young saw a pattern of alternating light and dark bands on the screen. He realized that this result is similar to that seen when two stones are thrown into water at the same time. A series of waves is sent out by each stone, and where the waves meet, they combine. Where two wave peaks overlap, there is a peak of double height. Where two troughs overlap, there is a deep trough. This is called interference of waves. Interference could happen in Young's experiment only if light were a form of

▲ Two stones dropped into a still pond produce circular patterns of spreading ripples. Where they meet, the two wave patterns interfere. Where the ripples cancel out, the result is still water. Where two crests combine, a large ripple results. Light behaves in the same way, but instead of ripples, light produces light and dark bands.

◄ Clear plastic rulers and protractors show colored patterns in some lights. The colors are caused by the plastic splitting light into different beams, which then interfere with each other.

► Colored patterns form in soap bubbles when light reflected from the top of the soap film interferes with light reflected from the bottom. The delicate, swirling rainbow colors seen in thin oil films on water are produced in the same way.

wave. With white light, Young's experiment produced a pattern of colored bands. This is because the different colors that make up white light have different-sized waves, which cancel out in different places.

The modern idea is that light waves consist of packets of light energy called photons. In some situations the photons are important, and light behaves like a stream of particles. In others the wave properties are more important, and light behaves like waves. So the two explanations by Newton and Huygens each provide a part of the answer to the question "What is light?"

► Interference patterns are produced when two beams of light, with their waves in step, overlap. Bright bands are seen where the waves reinforce each other. Dark bands are seen where the waves cancel out.

Color

The colors of light correspond to waves of differing lengths. Red light, for example, has a wavelength – the distance between successive crests of a wave – of about 700 nanometers. One nanometer is one billionth of a meter (about a billionth of a yard). Violet light has the shortest wavelength in visible light, about 400 nanometers. Yellow, green, and blue light have wavelengths between these values. White light is a mixture of light of all wavelengths.

When colored lights are shined together onto a white surface, the colors are added together. Any color can be produced by mixing combinations of red, green, and blue light. These are the three primary colors of light and when mixed in equal proportions produce white light. Two primaries, for example red and green, make a secondary color, yellow.

Colored objects or paints absorb, or subtract, certain colors from light and reflect the rest. Our eyes see the reflected light only, and so the object appears to be the color of the reflected light. For example, red paint absorbs the green and blue colors in white light, and reflects only the red light. The secondary colors of light, yellow, magenta, and cyan, are the primary colors of paint. Many colors can be made by mixing them; for example a mixture of magenta and yellow make red. All three mixed equally together make black.

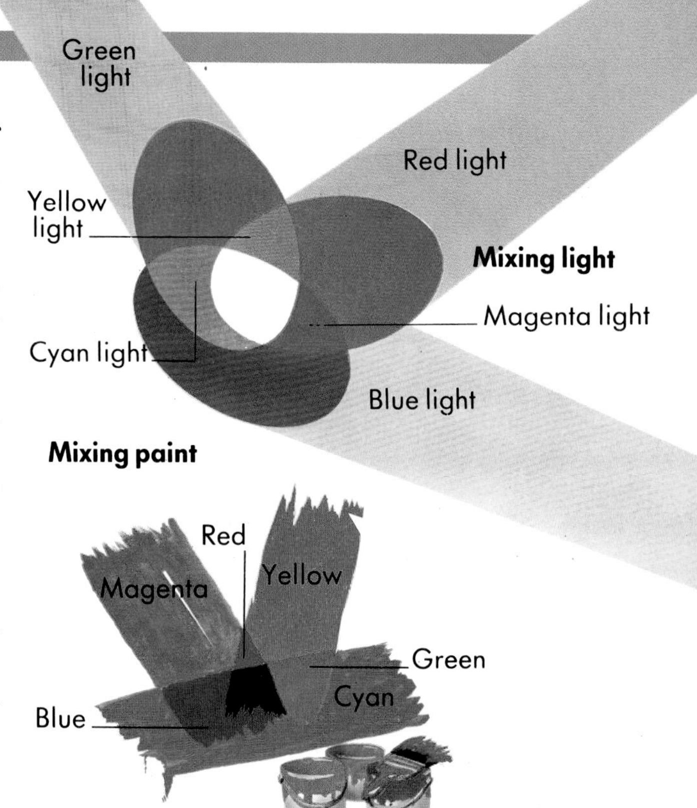

▲ Mixing paints is different from mixing colored lights, like those found in a theater or disco. Many colors of paint can be produced by mixing magenta, yellow, and cyan paints. With colored lights, any color can be produced by mixing red, green, and blue lights.

▼ The lights of a disco help to enhance the music and create an exciting atmosphere. The lights are usually wired to the loudspeakers so that they flash in time with the music.

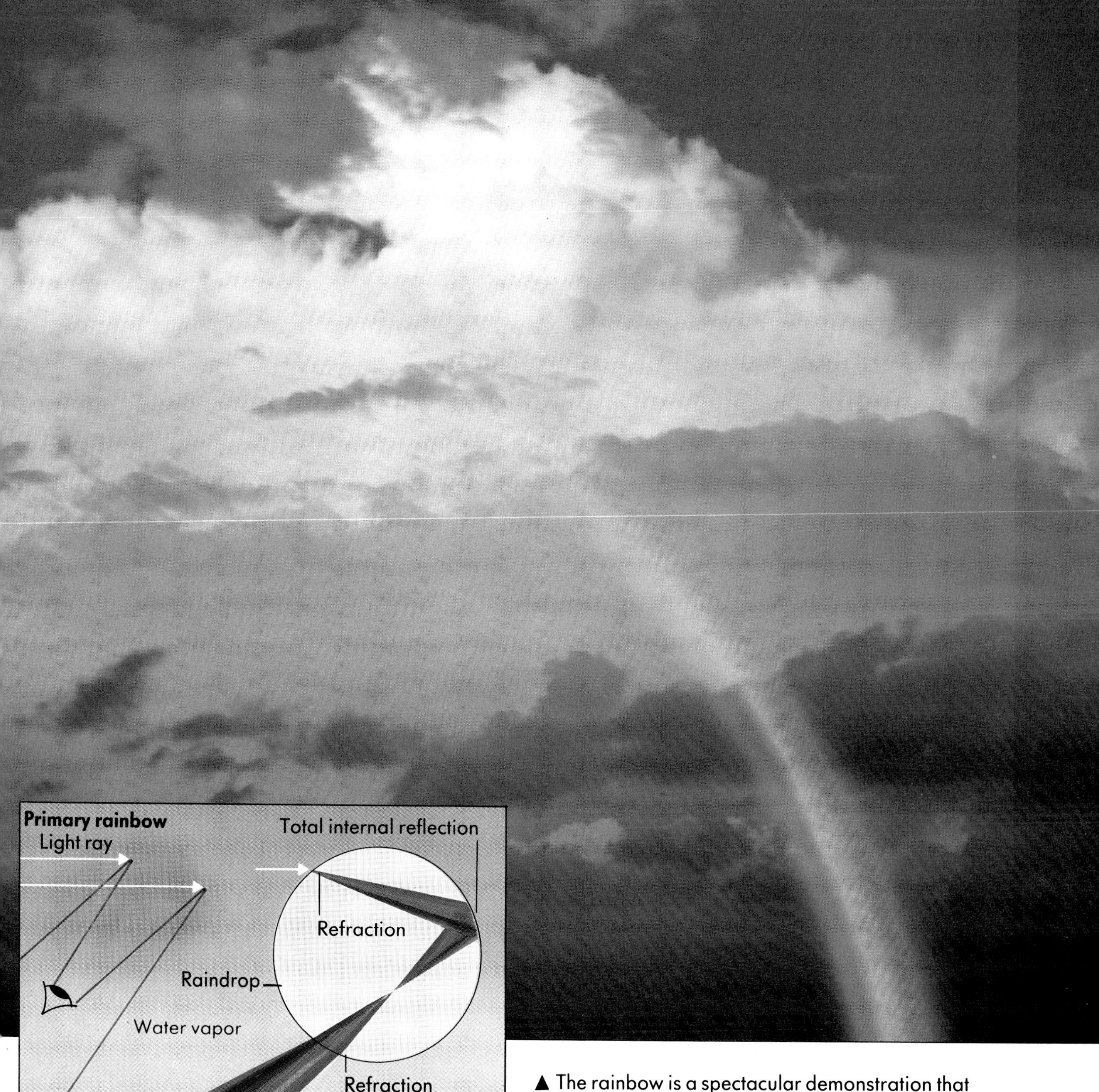

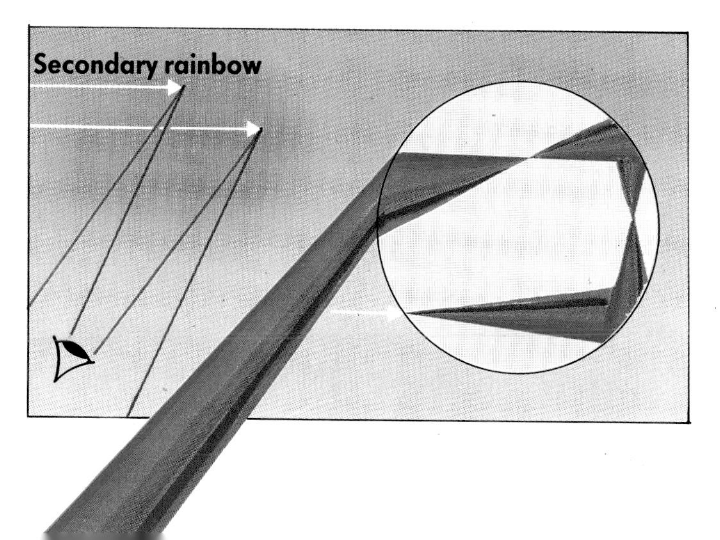

▲ The rainbow is a spectacular demonstration that white light is a mixture of colors. Rainbows occur when sunlight from behind the observer is refracted and reflected by water droplets in the air. Often a fainter "secondary" rainbow is seen outside the bright "primary" one. The colors in the secondary bow are in the reverse order to those in the primary.

◀ In forming the primary rainbow, light from the Sun is first refracted as it enters a raindrop, then instantly reflected from the back of the drop. Finally it emerges, spread into a band of colors. In forming the secondary bow, the light is reflected twice within the raindrop before it emerges. In the primary rainbow the light is reflected only once.

Lasers

An American, Theodore H. Maiman, invented the laser in 1960. A laser is a device that produces a very powerful beam of light of a single color. The word "laser" comes from a set of initials that stand for "light amplification by stimulated emission of radiation." This name was chosen because a laser persuades, or stimulates, its atoms to amplify, or make stronger, a flash of light.

Lasers are used in the home, factory, and hospital. A compact-disk player contains a low-power laser. The laser beam "reads" the music on the disk, just as the stylus reads the music on an ordinary record. Low-power lasers are also used at supermarket checkouts to read the bar codes on packages of food. Surgeons use high-power lasers to carry out delicate eye operations. In industry, lasers are used to cut and weld metal sheets. When the Apollo astronauts were on the Moon, they set up a mirror pointing at the Earth. Later, scientists shined a laser at the mirror. By measuring how long it took the laser beam to travel to the Moon and back, the scientists were able to measure the distance to the Moon very accurately.

Lasers are used to create three-dimensional photographs called holograms. To produce a hologram, laser light is shined on the object being photographed. The light is reflected off the object onto a photographic plate. At the same time, some light from the laser is shined directly onto the photographic plate. The two beams of light produce a complex pattern on the plate. Later, if laser light is shined through the plate, a three-dimensional picture is formed.

▼ Lasers are key research tools. They provide a source of light of a single wavelength, or pure color. Furthermore, the waves are all in step with each other. This greatly enhances the intensity of the light.

How a laser works

Electrons emit light when they lose energy. This happens when an electron jumps from one orbital to another one of less energy. As this happens, a photon is emitted.

When an electron in an atom is given extra energy, it is raised to a higher-energy orbital. The energized electron can then emit a photon and return to its usual orbital spontaneously. However, it can also be stimulated to fall back by a photon passing nearby. This is stimulated emission. The emitted photon and the passing photon move away in step, with their troughs and peaks matching exactly. The light produced in this way is said to be coherent.

The first lasers used a cylinder of ruby to produce their light. In the ruby laser a powerful flash tube is wrapped around the ruby cylinder. When the flash tube is turned on, the ruby becomes bathed in light. The ruby absorbs the light and its electrons move to high-energy orbitals.

One light photon is released, which then stimulates other energized electrons to release more photons. There is a rapid buildup of photons, until the light becomes bright enough to pass through a partly silvered mirror at one end. Ruby lasers are still used. Other types use gases or liquids.

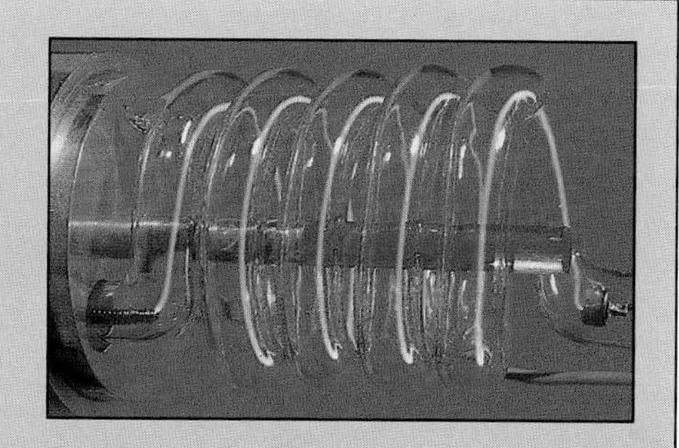

▲ The crystal and flash tube in a ruby laser. Crystals other than rubies have been developed for use in lasers. Most common are yttrium-aluminum-garnet (YAG) crystals, which allow the laser to operate continuously.

▼ When an atom absorbs light, electrons are raised to higher-energy orbitals. They can be stimulated to fall back to their usual orbital by a photon. When this happens, another photon is emitted, which moves away in step with the stimulating photon.

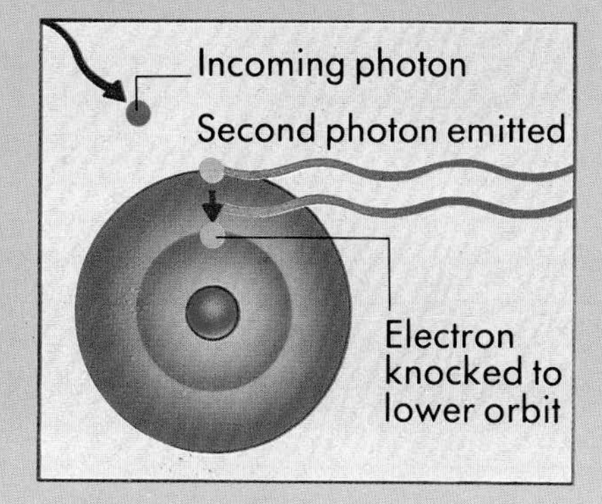

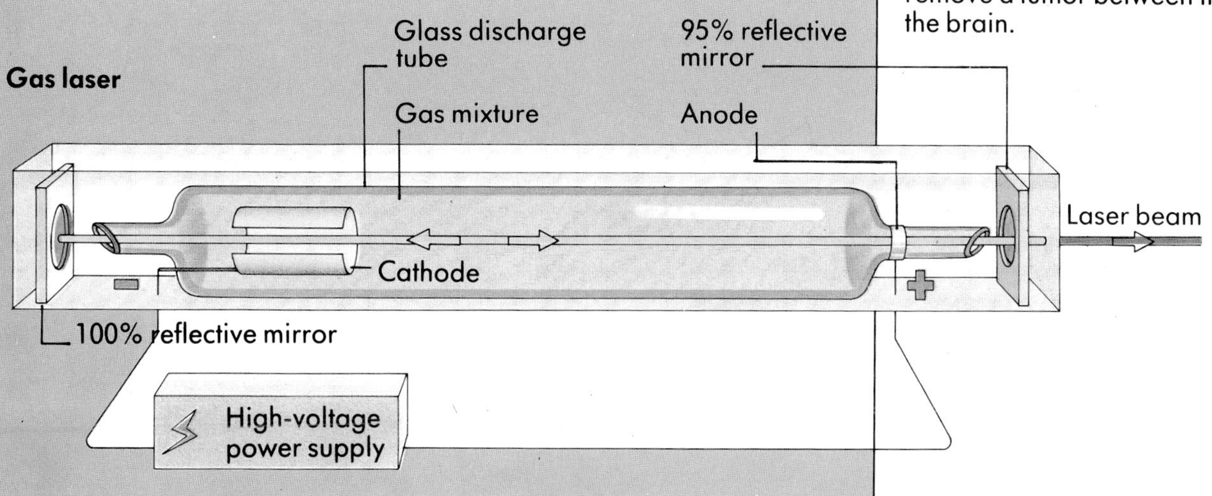

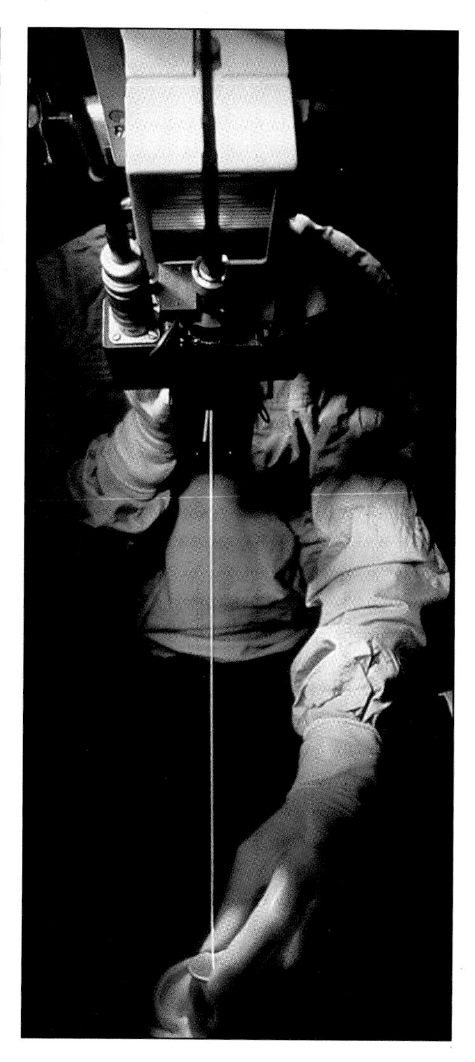

▲ Using a medical laser. Here a surgeon directs the beam from an argon laser through a small funnel into a patient's ear in order to remove a tumor between the ear and the brain.

The electromagnetic spectrum

What bees see

The eyes of animals and insects are often sensitive to wavelengths we cannot see. The bee responds to ultraviolet light. A flower that looks a uniform color to us, seems to have a dark center to a bee, enabling it to find the pollen.

In 1865 the Scottish physicist James Clerk Maxwell used mathematics to show that waves which were a combination of electricity and magnetism could spread through space. He called these waves electromagnetic waves.

You can visualize these waves by thinking about what happens when an electric charge is moved rapidly up and down. When still, the charge is surrounded by lines of electric force, which spread straight out from the charge. When the charge is moved up and down, the lines of force wiggle, in the same way that a stretched rope wiggles when one end is waved back and forth.

The wiggles in the lines of force move out from the charge in the same way that the wiggles move along the rope. However, the charge also generates a magnetic field as it moves, because it is a small electric current.

X-ray of a fractured leg

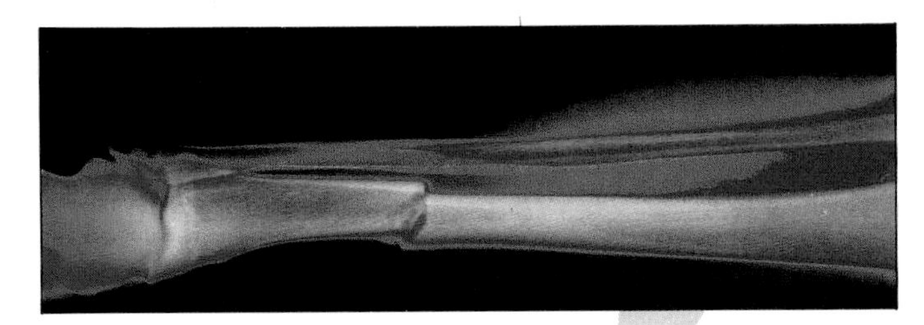

▼ The electromagnetic spectrum includes all forms of electromagnetic waves from gamma rays at the short-wavelength end to radio waves at the long-wavelength end. Visible light falls about the middle of the spectrum.

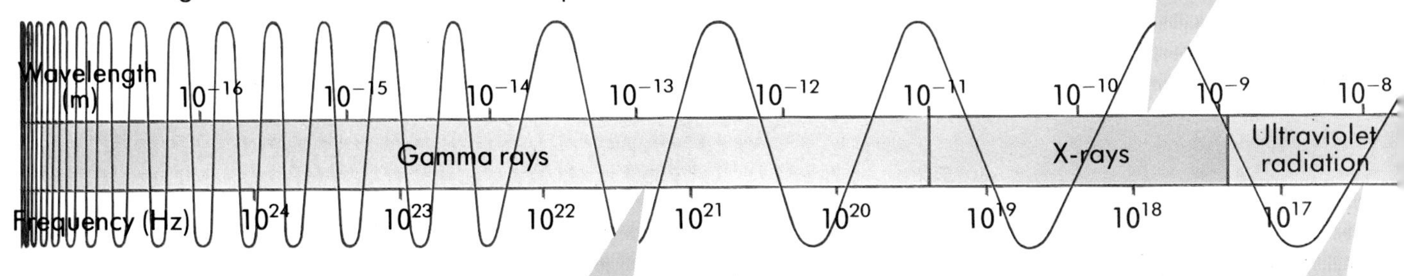

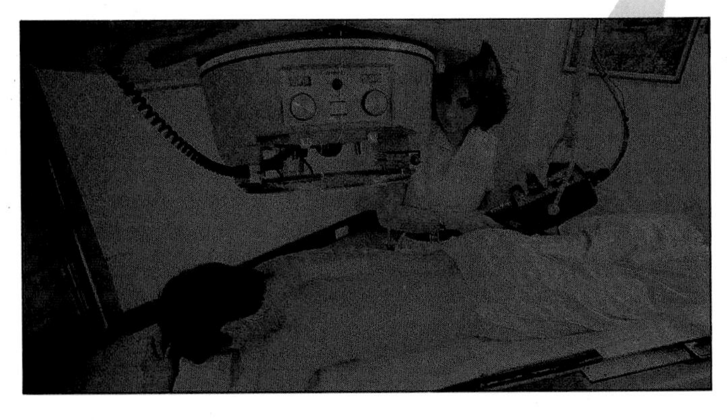

Gamma-ray therapy

An ultraviolet tanning bed

The lines of magnetic force form circles around the moving charge. They move out from the charge like ripples moving across a pond. They move outward at the same time and at the same speed as the electric wiggles. And so they form a wave that is a combination of changing electric and magnetic fields.

Maxwell calculated the speed of these electromagnetic waves to be the same as the speed of light, so he suggested that light consisted of electromagnetic waves. In 1889 the German physicist Heinrich Hertz produced radio waves and showed that they were electromagnetic waves too. They differed from light waves only in having longer wavelengths, or smaller frequencies. We now know that gamma rays, X-rays, microwaves, and infrared rays are all electromagnetic waves. They make up the electromagnetic spectrum.

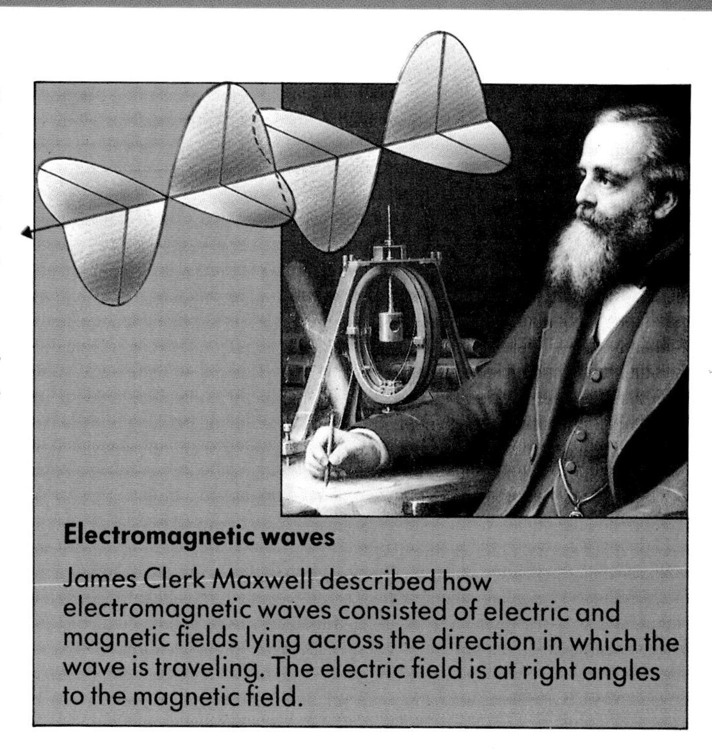

Electromagnetic waves

James Clerk Maxwell described how electromagnetic waves consisted of electric and magnetic fields lying across the direction in which the wave is traveling. The electric field is at right angles to the magnetic field.

Visible light fiber optics

A shortwave CB radio

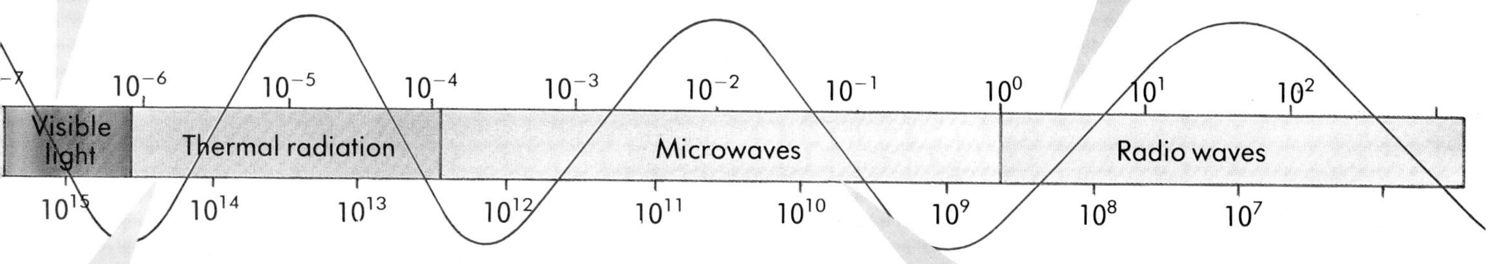

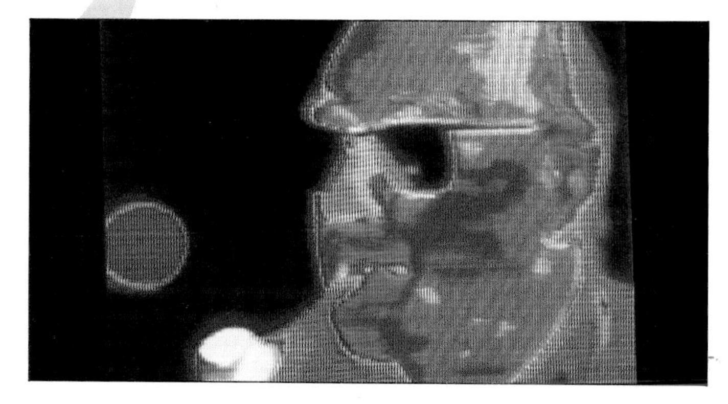

An infrared thermograph of a man smoking a pipe

A microwave radar antenna

Forces, energy, motion

Spot facts

- *On a 16th-century battleship, the guns were never fired all at once. This would have created such a big reaction force that the ship would have overturned.*
- *A flash of lightning releases up to 3 billion joules of energy. It takes 100 billion joules to send a rocket into space. A tropical hurricane releases about 100 quadrillion joules. Earthquakes can release up to 10 quintillion joules.*
- *If you can jump 1 m (about 1 yd.) high on Earth, you would be able to jump to a height of 6 m on the Moon.*

Imagine a ball being hit by a golf club. It is obvious that the force of the club on the ball starts the ball moving. But why does the ball continue to move after it has lost contact with the club?

The answer to this question can be found in the three laws of motion proposed by Isaac Newton in 1665. The three laws of motion explained how forces make objects move. These laws are still used today for tasks such as calculating the paths taken by spacecraft. Such calculations also make use of another great discovery by Newton: gravity. This is one of the great forces of the Universe; it attracts objects and makes them move.

► The Panavia Tornado, a European fighter-bomber. As a jet aircraft moves through the air, it is acted on by a number of forces. The force of gravity pulls it down. However, its wings provide an upward force, called lift, which keeps it in the air. The engines provide a forward force, which overcomes the resistance, or drag, of the air.

Force and movement

Nothing starts moving by itself. A push or a pull is needed to start any object moving. These pushes or pulls are called forces. When you kick a ball, the force which starts the ball moving is provided by your foot. As well as starting things moving, forces can stop moving objects.

Friction

When a ball rolls across the ground, a force called friction acts on it, and eventually stops it from moving. Without friction, the ball would go on moving at the same speed and in the same direction for ever. If a football bounces off a wall, the ball changes the direction in which it is moving because of the force exerted by the wall. Forces can speed up, slow down, or change the direction of a moving object.

A seat on a fairground carousel is continually changing its direction, so there must be a force acting on it. This force acts through the chain that holds the seat to the carousel. If the chain were to break, this force would cease to act. The seat would fly off and continue in a straight line. Any force that produces circular motion is called a centripetal force. Centripetal forces act towards the center of the circle.

Scientists believe that there are only four basic types of force. One is the electrical and magnetic force and another is gravity. The two other types of force, called the weak and the strong forces, are found only inside the atomic nucleus. All other forces are derived from these basic four.

▼ The people on this carousel feel a centripetal force, acting toward the center of their circular path. This force is a combination of their weight and the tension in the chain holding the chair. Confusingly, it is popularly called centrifugal force, which means one acting away from the center.

Laws of motion

In 1687 the English scientist Isaac Newton set down three laws of motion, which show how forces affect moving bodies. The first law says that an object at rest will stay at rest unless a force acts on it, and an object moving at a constant speed in a straight line will continue at the same speed and in the same direction unless a force acts.

The second law says that when a force acts on an object, the object changes its speed or direction of motion in the same direction as the force that has been applied. The change of speed or direction is called the acceleration of the object. The greater the force acting on a body, the greater the acceleration produced. The greater the mass of the body, the greater the force required to move it or change its direction.

The third law says that for every force, there is an equal force acting in the opposite direction. Newton illustrated this law with the example of a horse pulling a stone tied by a rope. While a forward force acts on the stone, the horse feels an equal force backward.

◀ The launching of the space shuttle illustrates all of Newton's laws of motion. As predicted by the first law, before the motor fires, the shuttle is still. In line with the second law, with the motors firing, the shuttle lifts off the launch pad. The third law indicates that the upward force on the shuttle is equal and opposite to the force acting on the hot gases streaming from the engines.

▼ Some dragster engines deliver an enormous force, accelerating the car to speeds of over 400 km/h (250 mph) in as little as 6 seconds.

Using these laws, scientists are able to explain how objects move and what happens when they collide. The first law explains a simple party trick. If a tablecloth is pulled quickly and firmly enough, it can be taken off the table without disturbing the dishes on it. The point is that the dishes do not experience the force that pulls away the cloth, and so they stay undisturbed. Unfortunately, as many tricksters have discovered at their own expense, there is another force which should be taken into account, friction. Friction is a force which acts when two surfaces rub together. It slows down the movement and may cause the dishes to move and crash to the floor.

One result of the second and third laws is that when objects collide, their total momentum does not change. The momentum of a moving object depends upon both the speed and the mass of the object. When a car is moving at a high speed, it has more momentum than when it is moving at a slower speed. Also, a big truck has more momentum than a small car moving at the same speed.

Sometimes if a moving ball in a pool or snooker game collides with a similar but unmoving ball, the first one stops. The second ball moves off with the same speed that the first ball had before the collision. The momentum of the first ball has been completely transferred to the second ball. But the total momentum after the collision is the same as the momentum before the collision. When you jump up and down, you are like a small ball banging into a very large ball, the Earth. After the collision, you stop moving, and the Earth absorbs your momentum. You make the Earth move. But because the Earth is 100 sextillion times heavier than you are, the movement is very tiny and you will not notice it.

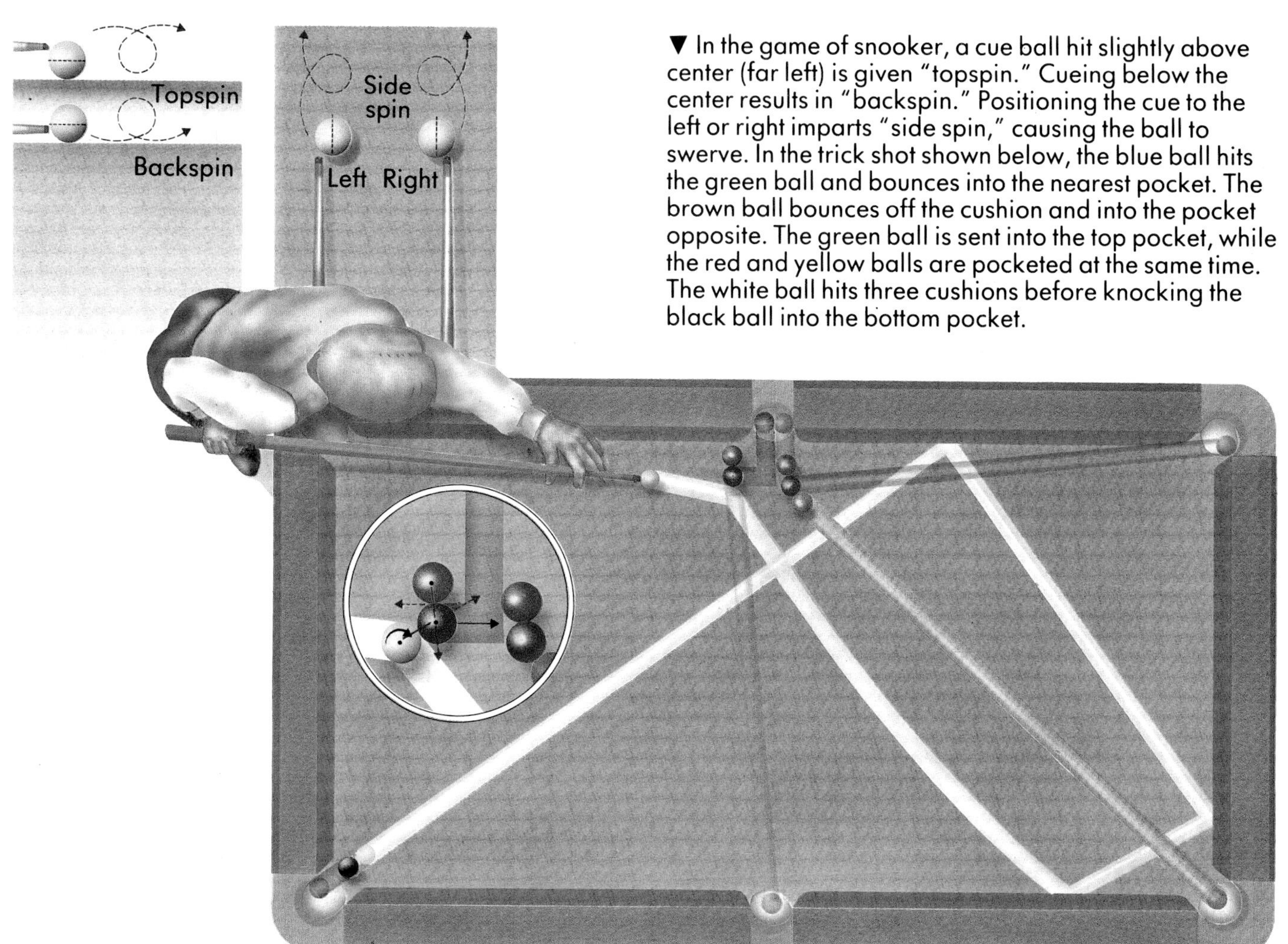

▼ In the game of snooker, a cue ball hit slightly above center (far left) is given "topspin." Cueing below the center results in "backspin." Positioning the cue to the left or right imparts "side spin," causing the ball to swerve. In the trick shot shown below, the blue ball hits the green ball and bounces into the nearest pocket. The brown ball bounces off the cushion and into the pocket opposite. The green ball is sent into the top pocket, while the red and yellow balls are pocketed at the same time. The white ball hits three cushions before knocking the black ball into the bottom pocket.

Energy and work

What is energy? We cannot touch it, see it, or weigh it. However, its effects are sometimes very obvious. When a lightning flash strikes, a bomb explodes, or a speeding train rushes by, it is clear that much energy is being used. The movement of the train shows one of the effects of energy. Nothing can move without energy. And because doing work involves movement, no work can be done without energy.

To a scientist, work is done whenever a force moves something. The greater the distance moved, and the greater the force involved, the more work is done. The more work done, the more energy is used. Work and energy are measured in units called joules, named after the British scientist James Prescott Joule, who lived in the 19th century. He did experiments to measure the heating effect of friction.

One joule is the work done when a force of one newton moves through a distance of one meter (about 1 yd.) A newton is a force that gives a mass of one kilogram (2⅕ lb.) the acceleration of one meter per second per second. This is equivalent to lifting a bag of sugar from one shelf to another in a cupboard.

It is clear that energy comes in different forms. Some forms of energy are obvious, others are more difficult to spot. A wound-up clock spring has energy because it can make the hands of the clock move. This form of energy is called stored, or potential, energy. Moving objects also have energy, called kinetic energy, because of their movement. Food contains energy in chemical form, which allows children to run about and play.

The different forms of energy can interchange. Electrical energy changes into heat energy in an electric fire. The chemical energy in food changes into energy of movement or heat in our bodies. The kinetic energy of the wind can be changed into electrical energy using a wind-driven generator. But when one form of energy changes into another, the total amount of energy remains the same. This is a statement of the law of conservation of energy.

▼ In 1977 a solar car drove more than 3,000 km (1,800 mi.) across Australia in six days. The car had 7,200 solar cells arranged around its body, which converted energy in sunlight into electrical energy to power electric motors to drive the wheels. The top speed was 72 km/h (45 mph).

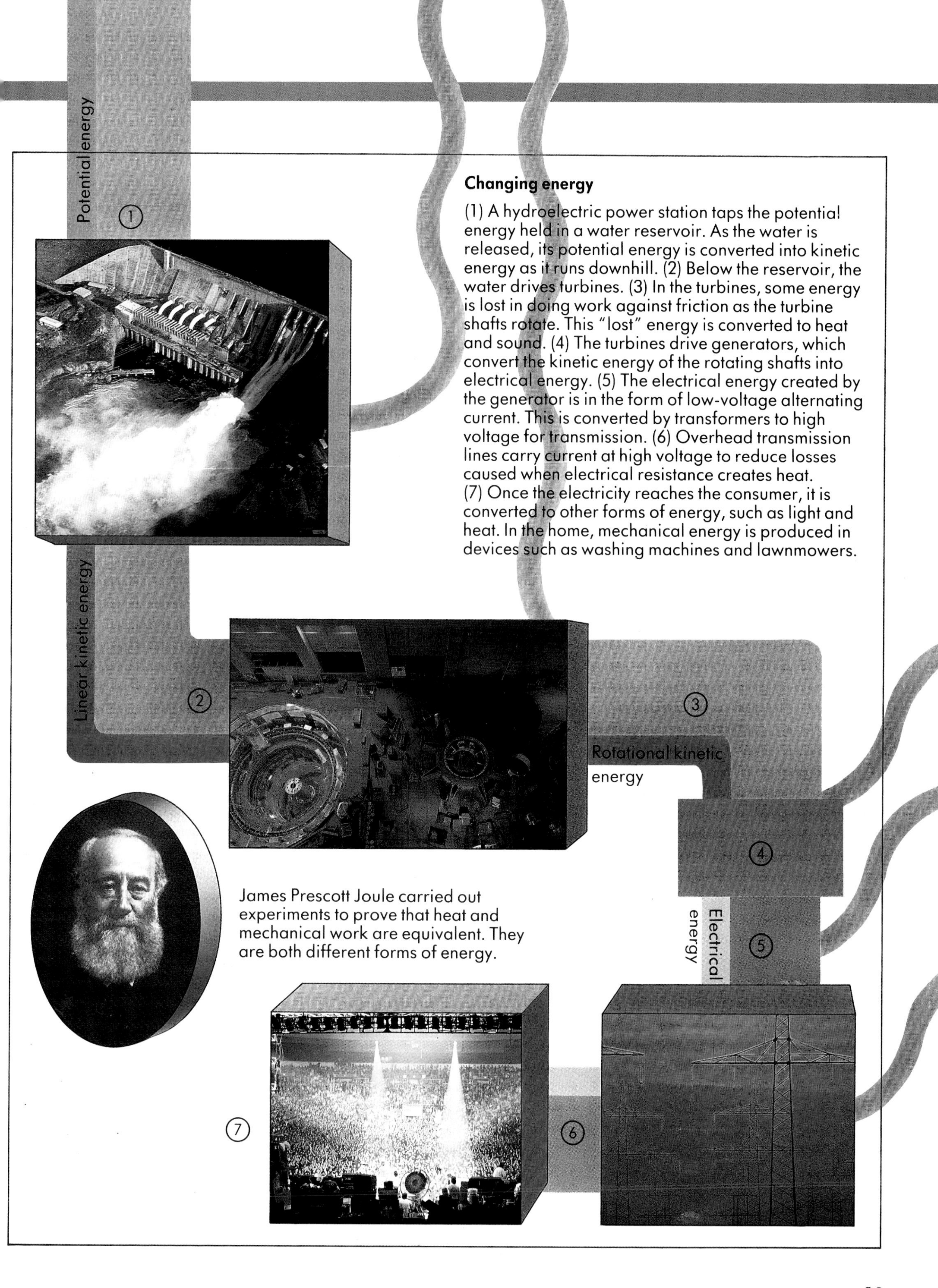

Changing energy

(1) A hydroelectric power station taps the potential energy held in a water reservoir. As the water is released, its potential energy is converted into kinetic energy as it runs downhill. (2) Below the reservoir, the water drives turbines. (3) In the turbines, some energy is lost in doing work against friction as the turbine shafts rotate. This "lost" energy is converted to heat and sound. (4) The turbines drive generators, which convert the kinetic energy of the rotating shafts into electrical energy. (5) The electrical energy created by the generator is in the form of low-voltage alternating current. This is converted by transformers to high voltage for transmission. (6) Overhead transmission lines carry current at high voltage to reduce losses caused when electrical resistance creates heat. (7) Once the electricity reaches the consumer, it is converted to other forms of energy, such as light and heat. In the home, mechanical energy is produced in devices such as washing machines and lawnmowers.

James Prescott Joule carried out experiments to prove that heat and mechanical work are equivalent. They are both different forms of energy.

Gravity

The first person to realize why things fall to the ground was Isaac Newton. It is said that he was sitting in the orchard of his house at Woolsthorpe, in Lincolnshire, England, in 1665. He saw an apple fall from a tree. He realized that the Earth must be pulling, or attracting, the apple. He went on to discover that all objects attract each other. The attractive force between objects is called gravity. You can feel this force if you try to lift anything. The weight of an object is due to the force of gravity between the object and the Earth.

Newton realized that gravity was a long-range force. The force of gravity of the Earth reaches beyond the Moon. It stops the Moon from flying off into space. In turn, the Moon's gravity pulls the Earth's seas toward it, causing tides. The force of the Sun's gravity reaches far out into space and controls the movements of the planets.

Newton's studies of gravity revealed that the force of gravity gradually got weaker away from the Earth. A person is attracted less in a high-flying aircraft than on the ground. However, the change in weight is very small and we do not notice it. At 25,000 km (15,000 mi.) above the Earth, you would weigh only about one-tenth what you do on the ground.

The force of gravity is smaller on the Moon than on Earth. This is because the force of gravity depends upon the amount of matter in the objects being attracted together. The more matter, or mass, the objects have, the stronger the force of gravity between them. Because the Moon has only one-sixth the mass of the Earth, its gravity is only one-sixth that of the Earth. A person who weighs 60 kg (120 lb.) on Earth would only weigh 10 kg (20 lb.) on the Moon. Astronauts can throw things much farther on the Moon because of the weak gravity.

◀ An astronaut in orbit above the Earth appears to be floating motionless in space. But gravity has not ceased acting. The astronaut is falling freely and moving forward at great speed. The combination of the two movements produces a circular path that keeps the astronaut at the same distance above the Earth.

▼ According to modern ideas, the gravitational force of a large collection of stars, such as a galaxy, bends the space around it. This causes light rays to bend rather than follow a straight path. If a bright object, such as a quasar, is behind the galaxy, two slightly separated images of the quasar can be seen from Earth. One image is the direct view of the quasar; the other is due to the light bent around the galaxy. This effect is called a gravitational lens.

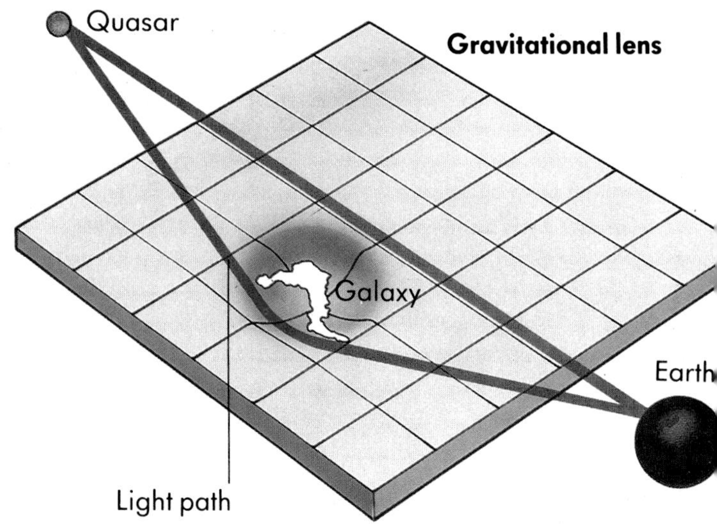

▲ Sky divers experience a force due to air resistance as they fall. This increases with speed, and at a certain speed becomes equal to the force of gravity, which is accelerating them downward. When this happens, the divers fall at a constant speed, called the terminal velocity. Sky divers can reach a speed of 298 km/h (185 mph) falling headfirst in the lower atmosphere.

◄ The Italian scientist Galileo began the scientific study of moving objects about 1590. He is said to have dropped objects of different weights from the Leaning Tower of Pisa, to show that all objects fall at the same rate. He also made many important astronomical discoveries using a telescope, which he constructed in 1609. He was the first to see the moons circling the planet Jupiter.

Galileo's experiment

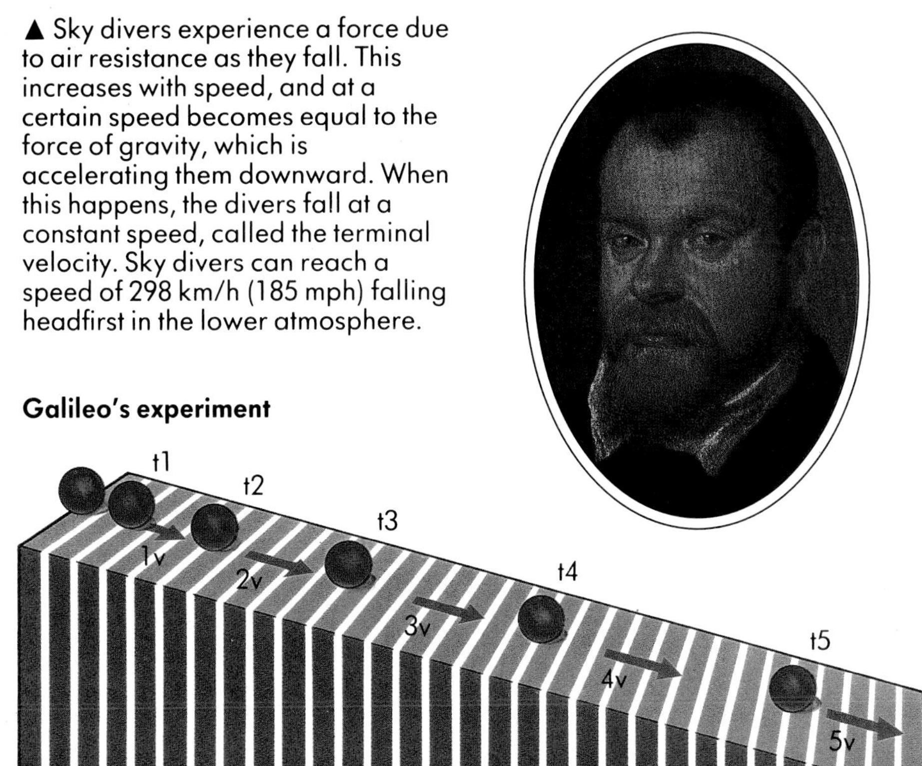

◄ An experiment performed by Galileo involved rolling balls down a gently sloping plank and measuring the distance moved in equal intervals of time (t). Unfortunately, Galileo did not possess an accurate clock; he used a water clock. Nevertheless, he was able to show that the speed or velocity (v) increased steadily as the ball moved down the slope. In other words, the force of gravity produced a steady acceleration on the ball.

Units of measurement

Units of measurement

This encyclopedia gives measurements in metric units, which are commonly used in science. Approximate equivalents in traditional American units, sometimes called U.S. customary units, are also given in the text, in parentheses.

Some common metric and U.S. units

Here are some equivalents, accurate to parts per million. For many practical purposes rougher equivalents may be adequate, especially when the quantity being converted from one system to the other is known with an accuracy of just one or two digits. Equivalents marked with an asterisk (*) are exact.

Volume
1 cubic centimeter = 0.0610237 cubic inch
1 cubic meter = 35.3147 cubic feet
1 cubic meter = 1.30795 cubic yards
1 cubic kilometer = 0.239913 cubic mile

1 cubic inch = 16.3871 cubic centimeters
1 cubic foot = 0.0283168 cubic meter
1 cubic yard = 0.764555 cubic meter

Liquid measure
1 milliliter = 0.0338140 fluidounce
1 liter = 1.05669 quarts

1 fluidounce = 29.5735 milliliters
1 quart = 0.946353 liter

Mass and weight
1 gram = 0.0352740 ounce
1 kilogram = 2.20462 pounds
1 metric ton = 1.10231 short tons

1 ounce = 28.3495 grams
1 pound = 0.453592 kilogram
1 short ton = 0.907185 metric ton

Length
1 millimeter = 0.0393701 inch
1 centimeter = 0.393701 inch
1 meter = 3.28084 feet
1 meter = 1.09361 yards
1 kilometer = 0.621371 mile

1 inch = 2.54* centimeters
1 foot = 0.3048* meter
1 yard = 0.9144* meter
1 mile = 1.60934 kilometers

Area
1 square centimeter = 0.155000 square inch
1 square meter = 10.7639 square feet
1 square meter = 1.19599 square yards
1 square kilometer = 0.386102 square mile

1 square inch = 6.4516* square centimeters
1 square foot = 0.0929030 square meter
1 square yard = 0.836127 square meter
1 square mile = 2.58999 square kilometers

1 hectare = 2.47105 acres
1 acre = 0.404686 hectare

Temperature conversions

To convert temperatures in degrees Celsius to temperatures in degrees Fahrenheit, or vice versa, use these formulas:

Celsius Temperature = (Fahrenheit Temperature − 32) × 5/9
Fahrenheit Temperature = (Celsius Temperature × 9/5) + 32

Numbers and abbreviations

Numbers

Scientific measurements sometimes involve extremely large numbers. Scientists often express large numbers in a concise "exponential" form using powers of 10. The number one billion, or 1,000,000,000, if written in this form, would be 10^9; three billion, or 3,000,000,000, would be 3×10^9. The "exponent" 9 tells you that there are nine zeros following the 3. More complicated numbers can be written in this way by using decimals; for example, 3.756×10^9 is the same as 3,756,000,000.

Very small numbers – numbers close to zero – can be written in exponential form with a minus sign on the exponent. For example, one-billionth, which is 1/1,000,000,000 or 0.000000001, would be 10^{-9}. Here, the 9 in the exponent -9 tells you that, in the decimal form of the number, the 1 is in the ninth place to the right of the decimal point. Three-billionths, or 3/1,000,000,000, would be 3×10^{-9}; accordingly, 3.756×10^{-9} would mean 0.000000003756 (or 3.756/1,000,000,000).

Here are the American names of some powers of ten, and how they are written in numerals:

1 million (10^6)	1,000,000
1 billion (10^9)	1,000,000,000
1 trillion (10^{12})	1,000,000,000,000
1 quadrillion (10^{15})	1,000,000,000,000,000
1 quintillion (10^{18})	1,000,000,000,000,000,000
1 sextillion (10^{21})	1,000,000,000,000,000,000,000
1 septillion (10^{24})	1,000,000,000,000,000,000,000,000

Principal abbreviations used in the encyclopedia

°C	degrees Celsius
cc	cubic centimeter
cm	centimeter
cu.	cubic
d	days
°F	degrees Fahrenheit
fl. oz.	fluidounce
fps	feet per second
ft.	foot
g	gram
h	hour
Hz	hertz
in.	inch
K	kelvin (degree temperature)
kg	kilogram
l	liter
lb.	pound
m	meter
mi.	mile
ml	milliliter
mm	millimeter
mph	miles per hour
mps	miles per second
mya	millions of years ago
N	north
oz.	ounce
qt.	quart
s	second
S	south
sq.	square
V	volt
y	year
yd.	yard

Glossary

absolute zero The lowest temperature that can be reached, −273.15°C, −459.7°F, or 0K. At this temperature, all molecular motion ceases.

AC Short for **alternating current.**

acceleration The rate at which the velocity of a moving body changes.

acid A substance that can provide hydrogen ions for a chemical reaction. It combines with a base to form a salt and water.

acoustics The science of sound.

alkali A strong base that dissolves in water, for example, caustic soda.

alloy A mixture of metals, or of a metal and a nonmetal. Steel, for example, is an alloy made up of iron and carbon, together with varying amounts of other metals.

alpha particle One type of particle given off during radioactive decay. It is identical to the nucleus of the helium atom.

alternating current (AC) Electric current that travels first in one direction, then the other. The electricity supplied to our homes by power lines is AC.

anode A positive electrode, for example, of a battery or electron tube.

atom The basic unit from which all matter is made up. It is the smallest part of a chemical element that retains the characteristic properties of that element.

atomic number The number of protons in the nucleus of an atom.

base A substance that can accept hydrogen ions from an acid. It can be thought of as the opposite of an acid. A base combines with an acid to form a salt and water.

battery A device for producing electric current by chemical action. The term is also used to refer to other devices for producing electricity, such as solar and nuclear batteries.

beta particle One type of particle given off during radioactive decay. It is identical to an electron, and may be negatively charged (like an ordinary electron) or positively charged (like a positron).

boiling A change of state in which a liquid turns rapidly into a gas at a particular temperature, the boiling point. The boiling point depends on the external pressure.

bonding The means by which atoms link together to form chemical compounds, such as ionic, covalent, and metallic bonding.

Brownian motion The constant erratic motion of small particles in a liquid or gas because of collisions with the fast-moving molecules of the fluid.

capillarity A property of water and other liquids which makes them tend to rise or fall in narrow tubes. It is an effect of surface tension.

catalyst A substance that alters the speed of a chemical reaction, but remains itself chemically unchanged.

cathode A negative electrode, for example, in a battery or electron tube. In general it is a source of electrons.

cathode-ray tube An electron tube in which a beam of electrons is manipulated to form an image on a fluorescent screen. The picture tube in a TV set is a cathode-ray tube.

cell, electric A device that produces electricity by certain means. For example, a dry-cell battery produces electricity by chemical means; a solar cell converts the energy in sunlight into electricity.

center of gravity A point within a body or system at which the weight or mass of a body appears to be concentrated.

centripetal force A force on a body moving in a circle, directed toward the center of that circle, which keeps the body traveling in its circular motion.

change of state A change in the physical state of a substance, for example, from solid to liquid, or from liquid to gas.

chemistry The branch of science that studies the nature and properties of substances and the way they interact.

circuit, electric The path along which electric current flows.

compass, magnetic An instrument for finding direction which makes use of the Earth's magnetic field.

compound A substance that is made up of the atoms of more than one chemical element.

condensation The change of state when a gas changes back into a liquid.

conduction The passing on by a substance of electricity or heat (thermal conduction). In metals, which are good conductors, conduction is brought about by the flow of electrons.

convection The method by which heat flows through a gas or liquid. It involves the bodily movement of the fluid molecules.

covalent bonding A method of bonding in which combining atoms share electrons.

current, electric The flow of electricity in a conductor, being the flow of electrons. By convention current is said to flow from the positive electrode to the negative. But the electrons in reality flow in the opposite direction.

crystals Solids with naturally

formed flat faces. Minerals form distinctive crystals in cavities in rocks. The shape of a crystal reflects the regularity of its internal structure.

DC Short for **direct current.**

diffraction The spreading out of a wave (water, light, sound) after it passes through a narrow aperture or around an obstacle.

diffusion The gradual mixing of different gases or liquids when they are put together, brought about by the movement of their molecules.

direct current (DC) One-way electric current, like that produced by a battery.

electricity The effects brought about by the presence of positive and negative charged particles, and the flow of charged particles, usually electrons, through wires, gases, and so on.

electrolysis Producing a chemical reaction in a substance in solution or when molten by passing an electric current through it.

electromagnet A temporary magnet consisting of a coil of wire wound around an iron core. It is a magnet only while electric current is being passed through the coil.

electromagnetic radiation A kind of radiation that consists of electric and magnetic vibrations, which travel in the form of a wave. It comprises a family of waves, which differ from one another in their frequency and wavelength. These waves include X-rays, light rays, and radio waves.

electromagnetism The study of the close relationship between electricity and magnetism.

electron The smallest of the three main particles in atoms. It has a negative electric charge.

electroplating The coating of one metal on another by means of electrolysis.

elements, chemical Simple substances made up of atoms with the same atomic number. They are the building blocks of matter.

evaporation The escape of molecules from the surface of a liquid to form a vapor.

field The area in which a body exerts an influence, such as a magnetic field.

force Commonly a push or a pull; something that, when it acts on a body, tends to change its state of motion or deform it.

frequency Of a wave motion, the number of complete waveforms (cycles) that pass a given point in a certain time. It is measured in hertz, or cycles per second.

friction A force that opposes motion of one surface over another when the two surfaces are in contact with each other.

fusion The process of melting, in which a substance changes from the solid to the liquid state.

gamma rays Electromagnetic radiation of short wavelength. Gamma rays are often given off in radioactive decay.

gas One of the three main states of matter. The particles in a gas travel freely at very high speed. Contrast **solid** and **liquid.**

Geiger counter An instrument for detecting and measuring atomic radiation.

generator, electric A device that produces electricity by converting mechanical energy into electrical energy. It uses the principles of electromagnetism.

gravity (or gravitation) The force of attraction that exists between any two lumps of matter. It is one of the basic forces of the Universe.

half-life The time it takes half of a given amount of radioactive material to decay.

heat A common form of energy. It passes from one body to another when there is a temperature difference between them. Heat is a measure of the internal, kinetic energy possessed by particles.

hologram A three-dimensional picture (that is, one with depth) created by means of laser light.

inorganic chemistry One of the main branches of chemistry, concerned with the study of the chemical elements and their compounds, except carbon compounds containing hydrogen.

insulator A material that does not conduct electricity or heat well.

integrated circuit A complete electronic circuit, all of whose components are collected, or integrated, on a single piece of semiconductor material, usually silicon.

ion An atom or molecule that has lost or gained one or more electrons. Metals lose electrons to form positive ions, or cations; nonmetals gain electrons to form negative ions, or anions.

ionic bonding A form of chemical bonding that involves the transfer of electrons between combining atoms. Substances formed in this way are called ionic compounds.

interference The interaction between two similar waveforms. It generally results in either an increased or reduced amplitude of vibration.

isotopes Atoms of an element that have different numbers of neutrons in the nucleus.

kinetic energy Energy due to motion. Contrast **potential energy.**

kinetic theory The modern theory of matter, based on the idea that it is made up of particles (atoms or molecules), whose kinetic energy increases with temperature.

laser A device that produces an intense beam of parallel light of a single wavelength. The name stands for "light amplification by stimulated emission of radiation," which describes how laser light is produced.

latent heat The heat released or absorbed when a substance changes state.

lens A piece of glass or other transparent material with at least one curved surface. Lenses in optical instruments focus light rays.

liquid One of the three main states of matter. The particles in a liquid can move, but not independently. Contrast **solid** and **gas.**

magnetism The property, possessed most strongly by iron and a few other metals, of being able to attract similar materials. It is closely related to electricity.

mass A measure of the amount of matter in a body. Contrast **weight.**

matter The stuff of which the Universe is made up.

melting (or fusion) The change of state when a solid turns into a liquid. It takes place at a fixed temperature, the melting point.

metal An element that is typically dense, hard, tough, and shiny; that conducts heat and electricity well; and that can be hammered into a thin sheet or drawn into fine wire without breaking. About three-quarters of the chemical elements are metals, although not all of them have all of the above properties. One, mercury, is a liquid at ordinary temperatures.

metallic bonding The type of chemical bonding that occurs in metals. Most metals have a crystalline structure in which their atoms are packed closely together. They contribute their outer electrons to a common pool. The presence of these free electrons helps explain why metals are such good conductors.

molecule A basic unit of a compound in which the atoms are linked by chemical bonds, particularly covalent bonds. Molecules are the smallest units of a compound that have the typical properties of that compound.

momentum The product of the mass and the velocity of a body.

motor, electric A device that uses electrical energy to produce mechanical motion. It works on the principles of electromagnetism.

neutralization The reaction between an acid and a base, which results in a salt plus water.

neutron One of the three main particles in an atom. Neutrons are found, with protons, in the nucleus (except in nuclei of ordinary hydrogen). They have a similar mass to protons but have no electric charge.

noble gases The gases in Group 0 of the Periodic Table, which have full outer electron shells. They are chemically very inert, forming hardly any compounds with other elements.

nuclear fission The splitting of the nucleus of a heavy atom, such as uranium, which causes the release of abundant energy.

nuclear fusion The combining together of light atoms (particularly hydrogen) to form heavier ones, a process that releases enormous energy.

nucleus The central part of an atom, which contains most of its mass. The two main kinds of particles in the nucleus are protons and neutrons (except for an ordinary hydrogen nucleus, which has just a single proton).

Ohm's law When a direct current flows in a conductor, the voltage is proportional to the current.

organic chemistry The branch of chemistry concerned with the study of the wealth of carbon compounds containing hydrogen. Such compounds were originally termed "organic" because it was thought that they could be made only by living organisms.

oxidation A common type of chemical reaction, in which an atom or group of atoms loses electrons. It is always accompanied by reduction, in which an atom or group of atoms gains electrons. This type of reaction is often termed redox (reduction-oxidation).

Periodic Table An arrangement of the chemical elements in which they are presented in horizontal lines (periods) and vertical columns (groups). This arrangement brings out relationships between elements.

plasma Gas that is almost completely ionized – its atoms are split up into positive ions and electrons. Gas exists in this state – often called a fourth state of matter – in the searingly hot interior of stars.

potential energy The energy stored in a body because of its position. A ball resting on a table has potential energy.

pressure The force acting on a unit area of surface, measured in such units as newtons per square meter, kilograms per square centimeter, or pounds per square inch.

proton One of the two main particles in the nuclei of atoms. It has a positive electric charge.

quarks The basic subatomic particles from which all other particles are believed to be made up.

radiation Energy given off in the form of electromagnetic rays or subatomic particles.

radioactivity A process in which unstable elements break down, emitting radiation from their nuclei. They may emit alpha or beta particles or gamma rays.

rare gas An alternative name for **noble gas.**

reflection The bouncing back of waves (such as light and sound) from a surface.

refraction The bending of light rays that occurs when they pass from one medium into another.

resistance In general the property of a substance to oppose motion, as in air resistance. Electrical resistance in a conductor opposes the flow of electric current.

salt A compound formed when an acid reacts with a base. It is an ionic compound, in which the elements are present as ions. Common salt, sodium chloride, is the most familiar salt.

semiconductor A material that has properties between an electrical conductor and an insulator. Conduction is brought about in it by the presence of minute quantities of impurities. Silicon is the most widely used semiconductor. It is used in integrated circuits.

solid One of the three main states of matter. The particles in a solid are bonded rigidly together.

sound A wave motion transmitted in the form of physical vibrations which our ears or a special device can detect.

spectrum The spread of colour obtained when light is split up into its constituent wavelengths.

static electricity The electric charge that builds up on some materials when they are rubbed.

sublimation A change of state in which a solid turns into a gas without becoming a liquid first. All solids will sublime.

superconductor An electrical conductor that has lost all its resistance. Some metals and alloys become superconducting at very low temperatures, within a few degrees of absolute zero. Others do so at higher temperatures.

surface tension A force that exists at the surface of a liquid and that makes the liquid behave as if it had a skin.

temperature A measure of the hotness of a substance. In science temperatures are measured on either the Celsius (°C) or the Kelvin (K) scales.

transformer An electrical device that transforms, or alters, the voltage of an alternating current.

transistor An electronic device made of semiconductor material that is used in electronic circuits for such purposes as amplifying (strengthening) signals.

ultrasonic waves Sound waves that have a frequency higher than the human ear can detect.

vapor The gaseous state of a substance that is normally liquid or solid at ordinary temperatures; for example, water vapor.

vector A quantity having both magnitude and direction. Acceleration and velocity are vectors.

velocity The speed of a body in a certain direction.

voltage A measure of the potential difference, or electrical "pressure," in a circuit.

wavelength The distance between two successive crests or troughs of a wave motion (such as sound or light). For a given family of waves, such as electromagnetic waves, wavelength times frequency equals velocity. In the case of electromagnetic waves, this is the velocity of light.

weight The force experienced by a body due to gravity. In science weight is the product of mass and the acceleration due to gravity, and is measured in newtons. In everyday life, however, weight is usually expressed in units of mass, such as kilograms or pounds.

Index

Page numbers in *italics* refer to pictures. Users of this Index should note that explanations of many scientific terms can be found in the Glossary.

G

H

I

J

K

L

M

N

O

P

Q

R

S

T

U

V

W

X

Z

Further reading

Apfel, Necia H. *It's All Elementary: From Atoms to the Quantum World of Quarks, Leptons, and Gluons.* New York: Lothrop, Lee & Shepard Books, 1985.

Ardley, Neil. *Atoms and Energy.* New York: Warwick Press/Watts, 1982.

Ardley, Neil. *Discovering Electricity.* New York: Franklin Watts, 1984.

Ardley, Neil. *The World of the Atom.* New York: Gloucester Press/Watts, 1989.

Asimov, Isaac. *How Did We Find Out About Atoms?* New York: Walker and Company, 1976.

Asimov, Isaac. *How Did We Find Out About Superconductivity?* New York: Walker and Company, 1988.

Berger, Melvin. *Atoms, Molecules & Quarks.* New York: G.P. Putman's Sons, 1986.

Berger, Melvin. *Our Atomic World.* New York: Watts, 1989.

Berger, Melvin. *Solids, Liquids, and Gases: From Superconductors to the Ozone Layer.* New York: G.P. Putman's Sons, 1989.

Boyle, Desmond. *Energy.* Morristown, N.J.: Silver Burdett, 1982.

Cash, Terry, and Barbara Taylor. *Electricity and Magnets.* New York: Warwick Press/Watts, 1989.

Chaple, Glenn F., Jr. *Exploring With a Telescope.* New York: Franklin Watts, 1988.

Cooper, Alan. *Electricity.* Morristown, N.J.: Silver Burdett, 1983.

Fermi, Laura. *The Story of Atomic Energy.* New York: Random House, 1961.

Fleisher, Paul. *Secrets of the Universe.* New York: Atheneum, 1987.

French, P.M.W., and J.W. Taylor. *How Lasers Are Made.* New York: Facts on File, 1987.

Laithwaite, Eric. *Force: The Power Behind Movement.* New York: Franklin Watts, 1986.

Macaulay, David. *The Way Things Work.* Boston: Houghton Mifflin Company, 1988.

White, Jack R. *The Hidden World of Forces.* New York: Dodd, Mead & Company, 1987.

Whyman, Kathryn. *Electricity and Magnetism.* New York: Gloucester Press/Watts, 1986.

Picture Credits

b = bottom, t = top, c = center, l = left, r = right. APL Adams Picture Library, London. CD Chemical Designs, Oxford. ESA European Space Agency. FSP Frank Spooner Pictures, London. GSF Geoscience Features, Ashford, Kent. NHPA Natural History Photographic Agency, Ardingly, Sussex. OSF Oxford Scientific Films, Long Hanborough, Oxford. RHPL Robert Harding Picture Library, London. SCL Spectrum Colour Library, London. SPL Science Photo Library, London.

6 SPL/David Parker. 8 Images Colour Library, Leeds. 9 Spacecharts. 10l Imitor. 10tr Paul Brierley. 10br GSF. 11 Hutchison Library. 12bl Frank Lane Agency. 12tr APL. 12-13 J. L. Knill. 13 RHPL/J. Green. 14 Spacecharts. 15l Zefa. 15br Spacecharts. 15tr R. Kerrod. 18 SPL/Patrice Loiez, CERN. 19 University of Cambridge, Cavendish Laboratory. 20 SPL/Dr. Mitsuo Chtsuki. 21 (all pics) CD. 22t SPL/Powell, Fowler & Perkins. 22b University of Cambridge, Cavendish Laboratory. 23 SPL/Lawrence Berkeley Laboratory. 25t FSP/ Gamma. 25bl Susan Griggs Agency. 25br British Museum. 26 RHPL/Alan Carr. 27 SPL/U.S. Army. 28 Bridgeman Art Library. 29l Equinox Archive/ Institute of Archaeology. 29r GSF. 30t SCL. 30b Popperfoto. 30-31 British Coal. 31tl RHPL/Ian Griffiths. 31tr Zefa/Croxford. 32l SPL/Manfred Kage. 32r SPL/Prof. Edwin Mueller. 33 SPL/Dr. J. Burgess. 34,35 CD. 36 Zefa. 37l Topham Picture Source. 37tr Brian & Sally Shuel. 37cr CD. 38 SCL. 40 Spacecharts/R. Kerrod. 41t SPL/Dr. R. Clark & M. R. Goff. 41b APL/Gerard Fritz. 43 SCL/J. Bradbury. 44 NHPA/S. Dalton. 45l R. B. Mitson. 45r SPL/CNRI. 46 SPL/Jan Hinsch. 48 Vautier de Nanxe. 49 SPL/Lawrence Berkeley Laboratory. 50l SPL/John Howard. 50r Minolta. 51 Zefa/T. Ives. 52 SPL/David Taylor. 53 ESA. 54b Tim Woodcock. 54t Susan Griggs Agency. 55 RHPL. 56l Art Directors. 56r OSF/Manfred Kage. 57 SPL/David Parker. 58 SPL/U.S. Department of Energy. 59 SPL/Vaughan Fleming. 62l SPL/Alex Bartel. 62r SPL/Pacific Press Service. 63 Zefa. 65 Hutchison Library/R. Aberman. 66 SPL/David Parker. 67 Robin Kerrod. 68 Perkin Elmer. 69l John Watney. 69r Robin Kerrod. 70l Paul Brierley. 70r SPL/Martin Dohrn. 71t ACE Photo Agency/Jerome Yeats. 71b SPL/ Jeremy Burgess. 72 Zefa/K.L. Benser. 73 SPL/Phil Jude. 74 SPL/Hank Morgan. 75l Hughes Research Lab, Malibu. 75r SPL/Alex Tsiaras. 76t SPl/N. M. Tweedie. 76c SPL. 76bl SPL/Martin Dohrn. 76br SCL. 77t Institute of Electrical Engineers. 77ct SPL/David Parker. 77cr SCL. 77bl SPL. 77br SPL/ Martin Dohrn. 78 Jerry Young. 79 RHPL/A. Carr. 80t FSP. 80b SCL. 82 SPL/Andrew Clarke. 83t Zefa. 83c Hutchison Library/R. House. 83bl Bridgeman Art Library. 83bc David Redfern/ Stephen Morley. 83br Zefa/F. Damm. 84 Spacecharts/NASA. 85t ACE Photo Agency. 85b National Maritime Museum.